# Advances in Soil Science

# Advances in Soil Science

B.A. Stewart, Editor

# Advances in Soil Science

## Volume 3

Edited by B.A. Stewart

With Contributions by
Z. Gerstl, R.J. MacRae, G.R. Mehuys,
W.F. Spencer, R. Webster, R.E. White, and B. Yaron

With 58 Illustrations

Springer-Verlag
New York Berlin Heidelberg Tokyo

B.A. Stewart
USDA Conservation & Production Research Laboratory
Bushland, Texas 79012, U.S.A.

ISSN:0176-9340


Softcover reprint of the hardcover 1st edition 1985



Typeset by Ampersand Publisher Services, Inc., Rutland, Vermont.

9 8 7 6 5 4 3 2 1

ISBN-13: 978-1-4612-9559-4 e-ISBN-13: 978-1-4612-5090-6
DOI: 10.1007/978-1-4612-5090-6

# Preface

The world population in 1930 was 2 billion. It reached 3 billion in 1960, stands at 4.6 billion today, and is expected to reach 6 billion by the end of the century. The food and fiber needs of such a rapidly increasing population are enormous. One of the most basic resources, perhaps the most basic of all, for meeting those needs is the *Soil*. There is an urgent need to improve and protect this resource on which the future of mankind directly depends. We must not only learn how to use the soil to furnish our immediate needs, but also ensure that the ability of the soil to sustain food production in the future is unimpaired. This is indeed a mammoth task; a 1977 United Nations survey reported that almost one-fifth of the world's cropland is now being steadily degraded. The diversity of soil makes it necessary for research to be conducted in many locations. There are basic principles, however, that are universal. This series, *Advances in Soil Science*, presents clear and concise reviews in all areas of soil science for everyone interested in this basic resource and man's influence on it.

The purpose of series is to provide a forum for leading scientists to analyze and summarize the available scientific information on a subject, assessing its importance and identifying additional research needs. But most importantly, the contributors will develop principles that have practical applications to both developing and developed agricultures. It is not the purpose of the series to report new research findings because there are many excellent scientific journals for that need. Communications in scientific journals, however, are generally restricted to short and technical presentations. Therefore, *Advances in Soil Science* fills a gap between the scientific journals and the comprehensive reference books in which scientists can delve in depth on a particular subject relating to soil science.

The ultimate aim of the series is to stimulate action—action to determine where there are arable soils, action to develop technology for more efficient crop production on these soils, action to reduce the risk of degrading these soil resources, and action to determine on which soils our research efforts should be concentrated. Without such action, the task of producing adequate food in the future may simply be too great. By the time the world gets reasonably close to population stability, demand for food and other agricultural products could be three times present levels.

There are many audiences to reach. While intended primarily for scientists and students of soil science, this series also provides technical information for anyone interested in our natural resources and man's influence on these resources. The reviews are written by leading scientists from many countries, and will therefore provide the reader with information from a wide array of conditions. Such information will be particularly useful to professionals working in areas with developing agricultures because the reviews summarize and assess the significance of the technical literature.

B.A. Stewart

# Contents

# Contributors

Z. Gerstl, Institute of Soils and Water, The Volcani Center, Bet-Dagan 50-250, Israel

R.J. MacRae, Department of Renewable Resources, Macdonald College of McGill University, Quebec, Canada H9X 1C0

G.R. Mehuys, Macdonald College of McGill University, Quebec, Canada H9X 1C0

W.F. Spencer, U.S. Department of Agriculture, Agricultural Research Service, Soil and Environmental Science Department, University of California, Riverside, California 92521, U.S.A.

R. Webster, Rothamsted Experimental Station, Harpenden, Hertfordshire AL5 2JQ England

R.E. White, Department of Agricultural Science, University of Oxford, Oxford OX1 3PF, England

B. Yaron, Institute of Soils and Water, The Volcani Center, Bet-Dagan 50-250, Israel

# Quantitative Spatial Analysis of Soil in the Field

R. Webster*



*Rothamsted Experimental Station, Harpenden, Hertfordshire, AL5 2JQ, England


Advances in Soil Science, Volume 3

# I. Introduction

Soil scientists have recognized variation in soil from place to place for many years. They have portrayed the variation by dividing large regions into smaller parcels each of which is relatively homogeneous, and they have classified the soil to show similarities between soil in widely separated parcels. This procedure, which may be regarded as standard soil survey practice, requires appreciation of the scale of change, the abruptness or otherwise of change, the degree of correlation among different soil properties, and of relations in the landscape. Yet that appreciation has almost always been intuitive. Good soil surveyors have needed flair. Rarely have they gained their appreciation by quantitative analysis.

The standard procedure has undoubtedly been successful. It has also had its disappointments; for example, in small regions where there are no obvious boundaries and where the properties of interest are uncorrelated with visible change. Now, because soil scientists increasingly require quantitative estimates of soil properties for regions of varying sizes and wish to plan their surveys in the most economical way, standard procedure must be augmented by more rational and quantitative methods.

In other contexts, such as advisory work, field experimentation, and irrigation planning, soil scientists have been very concerned with sampling efficiency and the measurements of variation in soil. Beckett and Webster (1971) reviewed the extensive literature on this and the numerous attempts to plan sampling strategies. At that time, the only way to increase efficiency and the precision of estimates seemed to be to stratify the soil first and then to use standard sampling theory to determine the estimation variances. This was a simple combination of classical statistics and soil classification.

The limitations of this classical approach were clear, and some alternative way of representing variation was obviously desirable. Trend surface analysis—that is, a form of multiple regression with the spatial coordinates as independent variables—enjoyed a brief spell of popularity in geology, especially for oil exploration (Harbaugh and Merriam, 1968). There were a few attempts to apply it to soil. Walker *et al.* (1968) had some success using polynomials over distances of up to 100 m on some segments of landscape but not others. In general, however, there was no obvious functional relation between geographic position and soil and no theory from which to predict one. Soil properties behaved much more like random variables. Time series methods seemed much more promising, therefore, and Webster (1973, 1977), Webster and Cuanalo (1975), and Kozlovskii and Sorokina (1976) began to explore their value and to adapt them to the spatial distribution of soil, as had Matérn (1960) to forestry and Gandin (1965) to meteorology. As it happened, they were overtaken by events.

Matheron (1965), working in the mining context, had brought together a number of isolated results in spatial statistics into a coherent body of theory, the theory of regionalized variables. His thesis in French was followed by an English text (Matheron, 1971), but because of the highly mathematical treatment and unfamiliar setting it was still some time before earth scientists, and soil scientists in particular, saw in it what they had been seeking.

This work must be regarded as a breakthrough for soil science. Regionalized variable theory now provides the basis for describing spatial variation in soil quantitatively, for estimating soil properties and mapping them soundly, and for planning rational sampling schemes that make the best use of manpower. This review shows how each of these is accomplished using the theory with examples from the recent literature. It begins, however, with a description of a much earlier approach to the problem attributable to Youden and Mehlich (1937), one that deserved much more attention than it received and that still has a proper place in a survey scheme.

## II. Nested Sampling and Analysis

Nested or hierarchical sampling schemes are commonly used in survey research to distinguish variation deriving from two or more levels of subdivision of a population. In the present context, the soil of a region might be divided into districts, farms, and fields, and in a survey one might wish to measure the variation contributed by each to the total variation in some property over the whole region. By sampling the soil within fields and identifying the farm and district to which each sampled field belongs, the variance can be partitioned by a hierarchical analysis and components estimated. Youden and Mehlich (1937) adapted this technique to measure the variance associated with different spatial scales in Broome County, New York, as follows. On each of two soil series they chose nine primary stations (stage 1) approximately 1.6 km apart. At each station they chose two substations (stage 2) 305 m apart, and at each of these they chose two sampling areas (stage 3) 30.5 m apart, in each of which two sampling points (stage 4) were located 3.05 m apart. They thus had a complete balanced design, which ensured that the analysis was straightforward. Note also the geometric progression of the spacings. In one series, the Culvers series, a coarse-loamy, mixed, mesic Typic Fragidcherpt, a sample of topsoil (0 to 15 cm) was taken at each sampling point and its pH measured. In the other series, Sassafras, a fine-loamy, siliceous, mesic Typic Hapludult, the pH of the A and B horizons was measured on samples from each point. Youden and Mehlich then performed an analysis of variance on each set of measurements, and their paper presents the results in the usual tabular form.

The model of variation is

$$Z_{ijkl} = \mu + A_i + B_{ij} + C_{ijk} + \varepsilon_{ijkl}, \qquad [1]$$

where $z_{ijkl}$ is the pH at point $l$ in the $k$th sampling area in the $j$th substation in station $i$; $\mu$ is the mean of $z$ in the region; $A_i$, $B_{ij}$, and $C_{ijk}$ are random variables associated with stages 1, 2, and 3 with variances $\sigma_1^2$, $\sigma_2^2$, and $\sigma_3^2$, respectively; and $\varepsilon_{ijkl}$ is a residual term with variance $\sigma_4^2$. Table 1 shows the general form of the analysis of variance for a four-stage balanced design with the contributions to the degrees of freedom and mean squares. In this study $n_1 = 9$ and $n_2 = n_3 = n_4 = 2$, so it is a simple matter to compute the estimated components of variance $\sigma_1^2$, $\sigma_2^2$, $\sigma_3^2$, and $\sigma_4^2$. Table 2 gives these.

The results have several features. In the Culvers series, the largest component of variance, accounting for almost 40% of the total, derives from the largest spacing, 1.6 km. The variance in stage 4 from points only 3.05 m apart accounts for 20% and is by no means the smallest. Turning to the Sassafras series, we see that the 3.05 m spacing contributes even more variance, 30.2% and 41.5% in the A and B horizons, respectively. In the A horizon, the 1.6 km spacing contributes nothing to the variance, most of which derives from spacings between 30 and 300 m.

The results are best appreciated by accumulating the components and plotting them against sample spacing, as in Figure 1. The variance in all three increases with increase in sample spacing, but there the similarity

**Table 1.** Degrees of Freedom and Mean Squares for a Four-Stage Balanced Hierarchical Analysis of Variance

| Source (stage)[a] | Degrees of freedom | Components of mean square |
|---|---|---|
| 1 | $n_1 - 1$ | $\sigma_4^2 + n_4\sigma_3^2 + n_3 n_4 \sigma_2^2 + n_1 n_2 n_3 \sigma_1^2$ |
| 2 | $n_1(n_2 - 1)$ | $\sigma_4^2 + n_4\sigma_3^2 + n_3 n_4 \sigma_2^2$ |
| 3 | $n_1 n_2(n_3 - 1)$ | $\sigma_4^2 + n_4\sigma_3^2$ |
| 4 | $n_1 n_2 n_3(n_4 - 1)$ | $\sigma_4^2$ |
| Total | $n_1 n_2 n_3 n_4 - 1$ | |

[a]For each stage, $g$, $n_g$ is the number of subdivisions within stage $g - 1$ and $\sigma_g^2$ is its component of variance.

**Table 2.** Components of Variance of pH in Two Soil Series in Broome County, New York

| Stage | and spacing (m) | Degrees of freedom | Culvers series 0–15 cm | | Sassafras series A horizon | | Sassafras series B horizon | |
|---|---|---|---|---|---|---|---|---|
| | | | Estimated component | Percentage of variance | Estimated component | Percentage of variance | Estimated component | Percentage of variance |
| 1 | 1600 | 8 | 0.02819 | 39.7 | 0 | 0 | 0.00569 | 16.7 |
| 2 | 305 | 9 | 0.02340 | 32.9 | 0.04440 | 60.3 | 0.00366 | 10.8 |
| 3 | 30.5 | 18 | 0.00552 | 7.8 | 0.00698 | 9.5 | 0.01055 | 31.0 |
| 4 | 3.05 | 36 | 0.01391 | 19.6 | 0.02225 | 30.2 | 0.01409 | 41.5 |

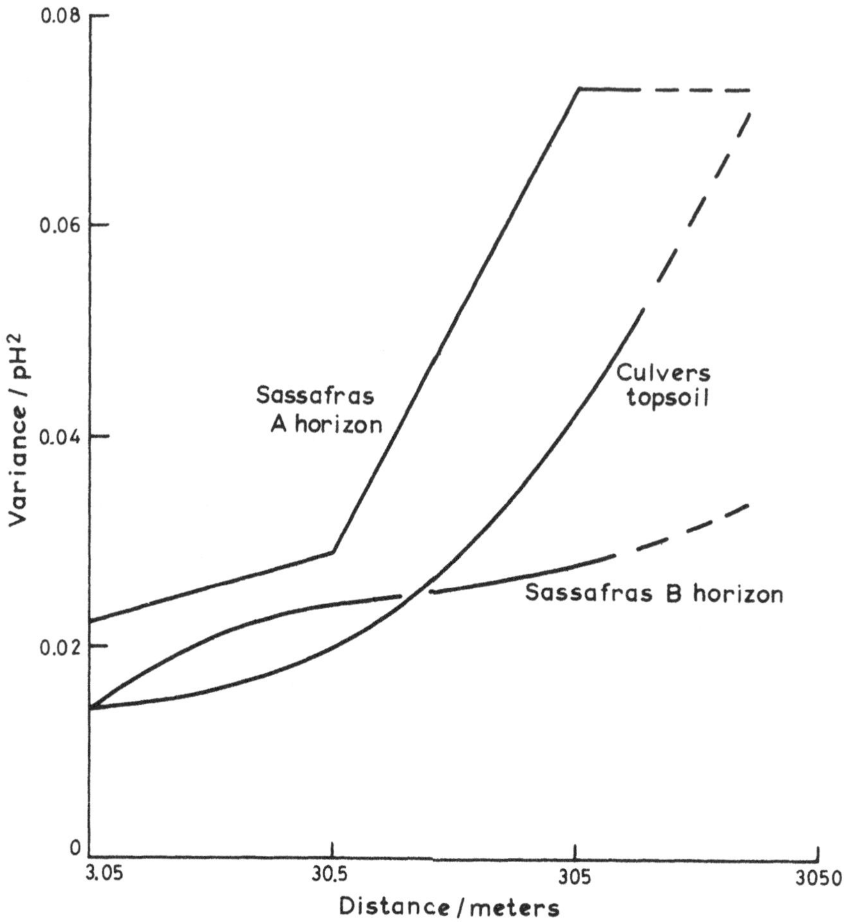


**Figure 1.** Graphs of accumulated variance of pH against separating distance of Broome County, New York.

ends. In the Culvers series, the variance appears to be increasing without limit, whereas in the A horizon of the Sassafras series, the variance has reached a maximum, which we now know as a *sill.* At the other end of the scale, two of the graphs appear to be leveling out to appreciable finite values, not zero, as might be expected. This we also know now to be quite general and recognize as the "nugget effect," which will be discussed in more detail later.

Two further examples illustrate the power of this technique. The first is from a study of soil variation on the 400 ha Ginninderra Experiment Station in the Australian Capital Territory (Webster and Butler, 1976). Again four sampling stages were chosen, eight primary stations at 180 m intervals with twofold subdivision at 50 m, 18 m, and 5 m, to give 64

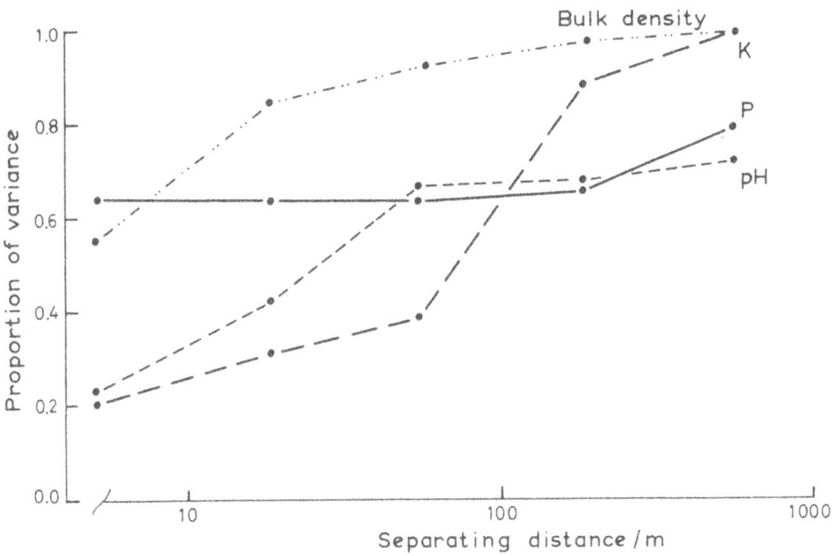


**Figure 2.** Graphs of accumulated variance of form topsoil properties against distance at Ginninderra, Australian Capital Territory.

sampling points at which several soil properties were measured on 10-cm diameter cores. Data for larger spacings were available, and these have been added in the analysis. The estimated components for four properties of the topsoil are accumulated and plotted against spacing in Figure 2. The most striking feature of these results is the difference in pattern between one soil property and another. For example, most of the variance in potassium content is contributed from spacings between 50 and 180 m, whereas for pH it derives from distances between 5 and 50 m. For bulk density most variance is present within 18 m, and for phosphorus it lies mainly within 5 m. The region was one for which it had been difficult to make a generally useful soil map, and these results show why.

The second example is from a study by Nortcliff (1978) in Norfolk, England. The sampling scheme has a less strict scale of distance. It was deliberately designed to distinguish major geological formations in the first stage, since it would have been foolish to have ignored such obvious features. Units in the second and third stages were chosen randomly within squares of 500 m and 100 m sides, respectively, rather than at constant spacings. The fourth and fifth stages were at 20 m and 5 m spacings. Nortcliff was especially interested in a strategy for general-purpose survey. He therefore recorded all the more obvious soil properties and transformed them to principal components before performing analysis of variance. Figure 3 presents the accumulated variances for the first six principal components. It distinguishes clearly the long-range variation of the leading

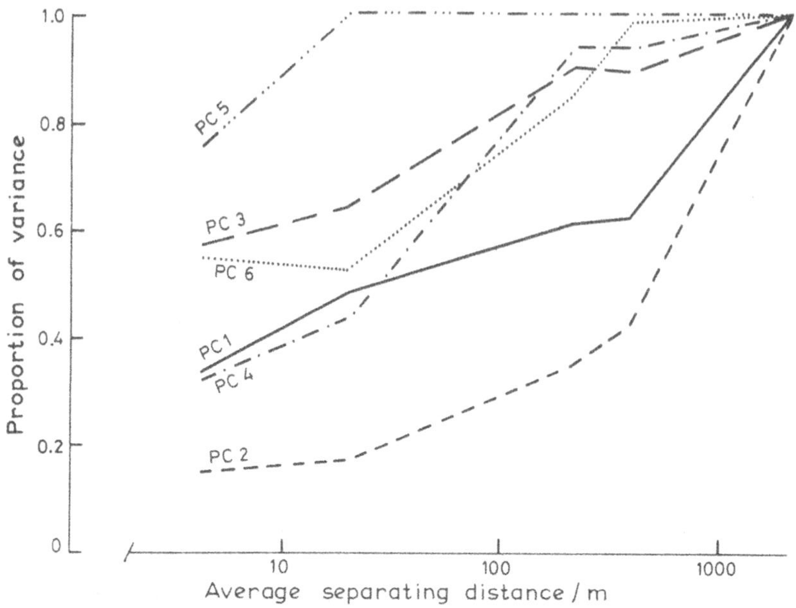


**Figure 3.** Graphs of accumulated variance of the six leading principal components of soil in Norfolk against distance. From Nortcliff (1978) with permission.

components from the dominantly short-range variation in the higher order ones. The correlated information seems to relate to the major geological differences and to be mappable at conventional medium mapping scales—1:25,000 or 1:50,000—whereas the uncorrelated information is not.

## III. Regionalized Variable Theory

In the nested analysis, variation is assumed to comprise a number of independent components, one for each spacing or stage of the hierarchy. If the spacings chosen increase in geometrical progression, this assumption is reasonable. Regionalized variable theory, however, takes a different approach. It considers differences between pairs of values of a property at places separated by any distance and expresses these as their variances. It also takes into account direction. Suppose we have the values $z(\mathbf{x})$ and $z(\mathbf{x} + \mathbf{h})$ at $\mathbf{x}$ and $\mathbf{x} + \mathbf{h}$, respectively, where $\mathbf{x}$ and $\mathbf{x} + \mathbf{h}$ are positions with one, two, or three spatial coordinates and $\mathbf{h}$ is a vector with both distance and direction, usually known as the *lag*, separating them. Then for this pair the variance per site is

$$s^2 = [z(\mathbf{x}) - \bar{z}]^2 + [z(\mathbf{x} + \mathbf{h}) - \bar{z}]^2, \qquad [2]$$

where $\bar{z}$ is the mean of the two values. Notice that $s^2$ is half the square of the difference:

$$s^2 = \tfrac{1}{2}[z(\mathbf{x}) - z(\mathbf{x} + \mathbf{h})]^2. \qquad [3]$$

Regionalized variable theory focused attention first on such differences and their variances. The quantity $s^2$ was therefore called the *semi-variance*, and the name has stuck. Nevertheless, it is the variance per site or observation. If further we have, say, $m$ pairs of observations separated by the same lag, $\mathbf{h}$, then we can define their average

$$\bar{s}^2 = \frac{1}{2m} \sum_{i=1}^{m} [z(\mathbf{x}_i) - z(\mathbf{x}_i + \mathbf{h})]^2. \qquad [4]$$

This is effectively what Youden and Mehlich did in computing the residual variance at 3.05-m spacing.

To make use of this simple notion and generalize equation [4], certain stationarity assumptions must be made. These are as follows.

1. The expected value of $z$ at any place $\mathbf{x}$ is the mean, $\mu$:

$$E[z(\mathbf{x})] = \mu. \qquad [5]$$

2. For any $\mathbf{h}$ the difference $[z(\mathbf{x}) - z(\mathbf{x} + \mathbf{h})]$ has a finite variance, which again is independent of $\mathbf{x}$:

$$\begin{aligned} \text{var}\,[z(\mathbf{x}) - z(\mathbf{x} + \mathbf{h})] &= E\{[z(\mathbf{x}) - z(\mathbf{x} + \mathbf{h})]^2\} \\ &= 2\,\gamma(\mathbf{h}). \end{aligned} \qquad [6]$$

These two assumptions constitute the *intrinsic hypothesis* of regionalized variable theory. They assume the following model of soil variation:

$$z(\mathbf{x}) = \mu_v + \varepsilon(\mathbf{x}), \qquad [7]$$

where $z(\mathbf{x})$ is the value of the property at position $\mathbf{x}$ within a region, $\mu_v$ is the mean value in that region, and $\varepsilon(\mathbf{x})$ is a spatially dependent random component with zero mean and variance defined by

$$\text{var}\,[\varepsilon(\mathbf{x}) - \varepsilon(\mathbf{x} + \mathbf{h})] = E\{[\varepsilon(\mathbf{x}) - \varepsilon(\mathbf{x} + \mathbf{h})]^2\} = 2\,\gamma(\mathbf{h}). \qquad [8]$$

In a large region, of course, we know that a soil property will vary from one part to another. Nevertheless, the property will commonly be locally stationary within some neighborhood $V$, and this condition is usually quite adequate for analysis in which $\mathbf{h}$ is limited to some maximum radius $r$ within which the relationships apply. It is for this reason that the subscript $V$ is used in equation [7].

Where the intrinsic hypothesis holds, we can expect the same degree of difference in the soil property at any two places $\mathbf{h}$ apart, whatever the actual values of the property are. In these circumstances, the sample value

$s^2$ in equation [4] is an unbiased estimate of the average semi-variance, $\gamma(\mathbf{h})$, in the population.

## A. Relation to Time Series Analysis

There is a close analogy between space and time, at least for one-dimensional space, and the methods of analysis are in large measure interchangeable. Traditionally, time series analysis has built on the auto-covariance and autocorrelation. The auto-covariance of a property that varies in time is defined as

$$\begin{aligned} C(h) &= \mathrm{E}\{[z(x) - \mu]\ [z(x + h) - \mu]\} \\ &= \mathrm{E}[z(x) \cdot z(x + h)] - \mu^2, \end{aligned} \qquad [9]$$

where $x$ and $x + h$ are points in time separated by a time lag $h$. The notation is changed slightly since $x$ and $h$ are now single valued in the one dimension. When $h = 0$, expression [9] defines the variance, i.e.,

$$\begin{aligned} C(0) &= \mathrm{E}[z^2(\mathbf{x})] - \mu^2 \\ &= \sigma^2. \end{aligned} \qquad [10]$$

The ratio $C(h)/C(0)$ is the autocorrelation, denoted by $\rho(h)$.

By combining equations [6], [7], and [8] the semi-variance is seen to be simply related to the auto-covariance and autocorrelation by

$$\begin{aligned} \gamma(h) &= C(0) - C(h) \\ &= \sigma^2\ [1 - \rho\ (h)]. \end{aligned} \qquad [11]$$

They are complementary. The more similar are values at lag $h$, the smaller is the semi-variance and the larger are the auto-covariance and autocorrelation. The latter must lie between 1 and $-1$.

To make use of this approach requires the assumption of full second-order stationarity. That is, both the mean and variance must be constant. It frequently happens in earth science that spatially distributed variables appear to have no finite *a priori* variance nor covariance. In these circumstances, the semi-variogram can exist without there being a corresponding covariance, and it is for this reason that the semi-variance rather than the covariance is used predominantly in spatial analysis.

## B. The Semi-Variogram and Its Estimation

Equations [6] and [8] define the semi-variance as a function of **h**, the lag. This function is the *semi-variogram*, $\gamma(\mathbf{h})$. In one dimension, $\gamma(h)$ can be estimated at regular intervals by sampling along transects. Thus, given a set of values $z(x_1)$, $z(x_2)$, . . . $z(x_n)$ we can estimate $\gamma(h)$, where $h$ is any integral multiple of the sampling interval, by

$$\hat{\gamma}(h) = \frac{1}{2(n-h)} \sum_{i=1}^{n-h} [z(x_i) - z(x_i + h)]^2. \qquad [12]$$

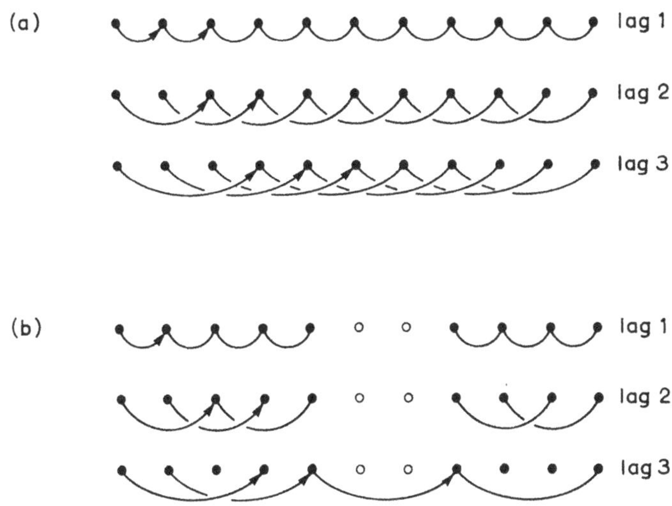


**Figure 4.** Comparisons for estimating semi-variances on linear transects at lags of 1, 2, and 3 sampling intervals, (a) for complete data and (b) where some observations are missing the open circles.

Figure 4a shows the comparisons involved for $h = 1, 2$, and 3. The result is an ordered set of values that constitute the *sample semi-variogram*. These can readily be plotted, and there are numerous examples now in the literature of soil science. Some are shown in Figures 6, 12, 13, 20, and 27.

This simple formula for estimating semi-variance is sensitive to extreme values of the differences $z(\mathbf{x}) - z(\mathbf{x} + \mathbf{h})$, especially as the squared difference follows a chi-squared distribution with one degree of freedom, which is highly skewed. Cressie and Hawkins (1980) investigated more robust estimators of $\gamma(\mathbf{h})$ and discovered that the fourth root of the usual squared difference,

$$y(\mathbf{x}) = \{[z(\mathbf{x}) - z(\mathbf{x} + \mathbf{h})]^2\}^{1/4}, \qquad [13]$$

had a distribution close to normal with negligible skew. They therefore used this formula to compute a mean $\bar{y}$ of $m$ separate $y(\mathbf{x})$. This value must, of course, be transformed back. Cressie and Hawkins showed that the expectation

$$\mathrm{E}[\bar{y}^4 / 2\,\gamma(\mathbf{h})] = 0.457 + \frac{0.494}{m} + \frac{0.045}{m^2}, \qquad [14]$$

and so the required back transformation gives the estimated semi-variance as

$$\hat{\gamma}(\mathbf{h}) = \bar{y}^4 / 2(0.457 + 0.494\, m^{-1} + 0.045\, m^{-2}). \qquad [15]$$

This is a recent development in variogram estimation, and there are no accounts yet of its being used for soil. Soil scientists should be aware of it, especially when analyzing data that are themselves skewed.

## C. Practical Problems

Formula [12] for the sample semi-variance assumes that there is a value for every sample point along the transect. More often than not there are deficiencies in the records that preclude such a tidy computation. The most common is that observations are lacking where the transect crosses roads, rivers, bare rock, and the like. The situation is then as illustrated in Figure 4b, and the summation is made over the actual number of pairs, $m < n - h$, that can be compared at lag $h$. In other instances several transects are surveyed because it is more convenient or because they cover the region more evenly than does just one. The semi-variance can again be computed by pooling the individual sums for each lag over all the transects and dividing by total $m$ comparisons.

A more serious difficulty arises where transects have been sampled at irregular intervals. This can be overcome by choosing a set of lags, $h_i$, $i = 1, 2, \ldots$ at arbitrary but constant intervals, $d$, and assigning to each a class with limits $(i - 1)d$ and $id$ and $h_i = (i - ½)d$ as its mid-point. Every pair of observations that are separated by the lag $h_i \pm d/2$ are then used to estimate $\gamma(h_i)$, and each pair contributes to one and only one estimate.

The effect of this procedure is to smooth the semi-variogram: The larger the increment the smoother will be the result. Some care is therefore needed to judge a suitable increment. If it is small there might be too few pairs of observations contributing to each semi-variance, so that the estimates are very imprecise. If, on the other hand, the increment is large then information can be lost by unnecessary smoothing. The best compromise will depend on the amount of data, the evenness of coverage, and the form of the underlying semi-variogram. Better still is to sample at regular intervals initially. The soil is accessible at most places, and there is rarely any good reason for not sampling systematically.

The confidence limits of semi-variances calculated in this way, whether from regular or irregular sampling, are uncertain. For a single transect, however, good practice is to have at least 100 sampling points and to estimate $\gamma(h)$ for lags up to no more than about one-fifth of the total run. This is a conservative guide; some workers are prepared to extend the lag to a third of the total length. If several transects are sampled then they should be long enough to provide at least 80 pairs for each estimate.

## D. Two Dimensions

In survey we are usually interested in the variation in a plane rather than in a single direction along a transect. The semi-variogram is then a two-dimensional function. A sample semi-variogram can be calculated quite straightforwardly if we have measured the soil at regular intervals on a two-dimensional grid. Suppose the grid has $m$ rows and $n$ columns. We estimate semi-variances as follows:

$$\hat{\gamma}(p, q) = \frac{1}{2(m-p)(n-q)} \sum_{i=1}^{m-p} \sum_{j=1}^{n-q} [z(i,j) - z(i+p, j+q)]^2 \qquad [16]$$

$$\hat{\gamma}(p, -q) = \frac{1}{2(m-p)(n-q)} \sum_{i=1}^{m-p} \sum_{j=q+1}^{n} [z(i,j) - z(i+p, j+q)]^2,$$

where $p$ and $q$ are the lags in the two dimensions. These equations enable half of the semi-variogram to be computed for lags from $-q$ to $q$ and from 0 to $p$. The semi-variogram is symmetrical about its center, so that if the full set of semi-variances is needed the remainder are readily obtained as

$$\hat{\gamma}(-p, q) = \hat{\gamma}(p, -q)$$

and

$$\hat{\gamma}(-p, -q) = \hat{\gamma}(p, q).$$

As in the one-dimensional case there will often be missing data, and, perhaps more likely, the region of interest will have an irregular shape. So again the quantity $(m - p)(n - q)$ in the denominators of equations [16] must be replaced by the actual number of paired comparisons in each sum.

The least tidy situation occurs where data are irregularly scattered in two dimensions. Every pair of observations is then separated by potentially different distances and directions. This difficulty is usually overcome by grouping the separations both in distance and direction. A series of lag distances and directions is chosen, usually to form regular progressions as in the one-dimensional case described above. A range in each is chosen, again usually equal to the class interval between successive lags, and applied so that the nominal lag lies at the center of the range. Each squared difference then contributes to the semi-variance for the lag class into which it falls by virtue of its actual separation. Figure 5 shows the geometry of the grouping. The nominal lag is represented by the line $OL$ of length $l$ and the direction $\theta$. The range in distance is $w$ and that in direction $\alpha$.

### 1. Representing Two-Dimensional Semi-Variograms

Sample semi-variograms in one dimension are readily drawn as in Figures 12, 20, and 27. If desired, the plotted points may be joined by straight lines, since the functions are continuous.

Displaying two-dimensional semi-variograms is not so straightforward since there are three dimensions involved: the two spatial dimensions plus that of the semi-variance. One way is to present the variogram as an

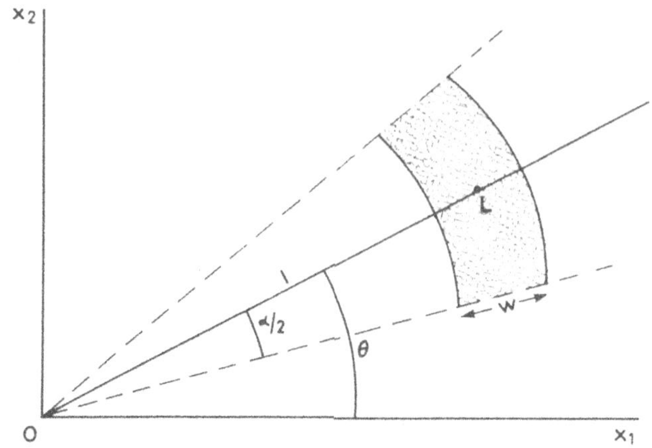


**Figure 5.** Grouping of lags by distance and direction. The striped area shows the extent of one group (see text).

isarithmic ("contour") map. Alternatively a perspective block diagram can be made of the interpolated surface. With a little more ingenuity stereograms can be drawn. As we shall see later, however, the forms of semivariogram in two dimensions are usually simple geometric extensions of one-dimensional forms, and a cylindrical projection of the sample values will often serve quite well.

To illustrate these possibilities we take as an example the study made by Burgess and Webster (1980a, 1980b) of the soil in Cae Ruel, one of the field at the Welsh Plant Breeding Station, Plas Gogerddan. The topsoil, 0 to 15 cm, had been sampled at 15.2 m intervals on a square grid. Several properties, including the stone content, were measured on bulked samples of 10 cores of 2.5-cm diameter chosen randomly within the 15.2 × 15.2 m squares around each grid node. There were approximately 450 observations in all.

In the original analysis semi-variograms were computed for four directions only: along the rows and columns of the grid and parallel to the two diagonals. Figure 6 shows the result. Using equations [16], however, a full two-dimensional semi-variogram can be computed, and as above, this can then be represented as an isarithmic map, Figure 7, block diagram, Figure 8, and cylindrical projection, Figure 9. The last shows a spread of points that expands away from the origin in a way similar to that in Figure 6. The semi-variances have been grouped into classes according to their angular separation. Each class, of which there are eight between 0 and 180 degrees, subtends 22.5 degrees, and is given a unique symbol. The significance of the oblique lines enveloping the points will be dealt with later.

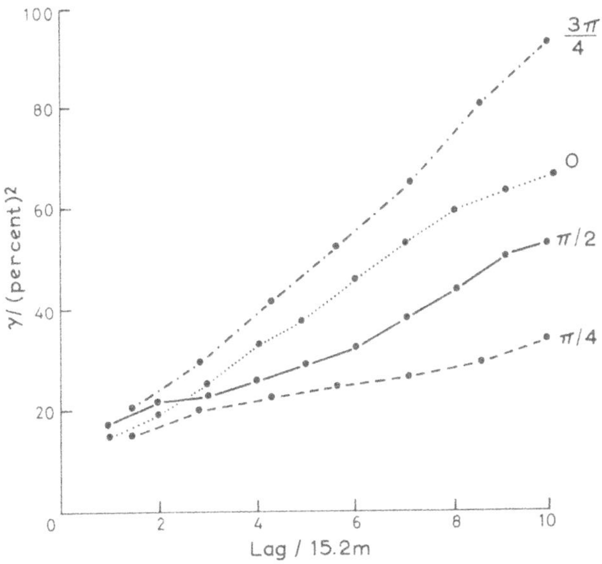


**Figure 6.** Sample semi-variograms of stone content in the topsoil at Plas Gogerddan in four principal directions.

## 2. Anisotropy

We note here that the semi-variogram of the stone content is not the same in all directions. Its gradient varies substantially from about 2 in direction $\pi/4$ to about 8 in the perpendicular direction. Thus, the stone content in this field varies anisotropically. This is not an unusual result. There are many situations in which anisotropic variation is recognized in soil survey. For example, where the land surface bevels a sequence of sediments the soil at points along the strike is more likely to be similar than that the same distance away in the direction of dip. On a point bar deposit or river levee we usually observe greater similarity in the soil if we travel parallel to the river than if we move at right angles to it. More generally it is common experience that variation is encountered more often when crossing the drainage lines of a region than when traversing along the contours. Thus an analysis of two-dimensional variation must allow for anisotropy.

### E. Size of Support

In material such as soil in which there is spatial dependence, the amount of variation present within a single core or block depends on the size of the block: The bigger its area or volume the more variation it embraces, and the less there is outside it. This affects the observed semi-variogram, and so a semi-variogram depends on the size, shape, and orientation of the individual samples on which measurements are made. Size, shape, and

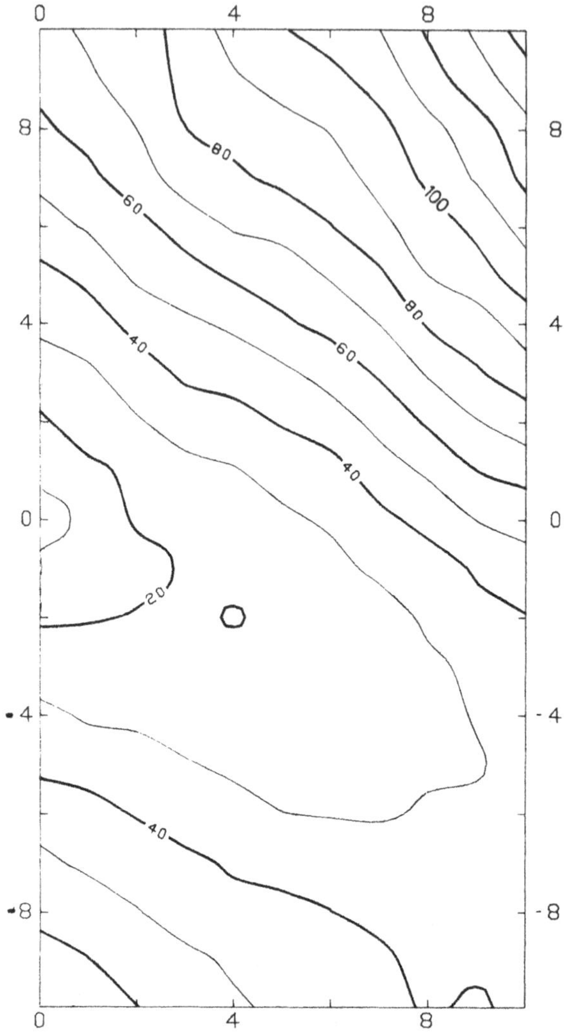


**Figure 7.** Isarithmic representation and sample semi-variogram of stone content at Plas Gogerddan. Border scales are in sampling intervals.

orientation constitute the *support* of the data. In general, measurements made on core samples encompass less variation than those made on larger volumes from pits. They in turn contain less than those made on bulked samples. Indeed, the aim of bulking is to diminish the variation between measurements by encompassing more within them. This increase in the support is known in geostatistics as *regularization*. Analysts should therefore realize that their results refer specifically to the particular support on which the measurements were made, and they should state the support when reporting their results.

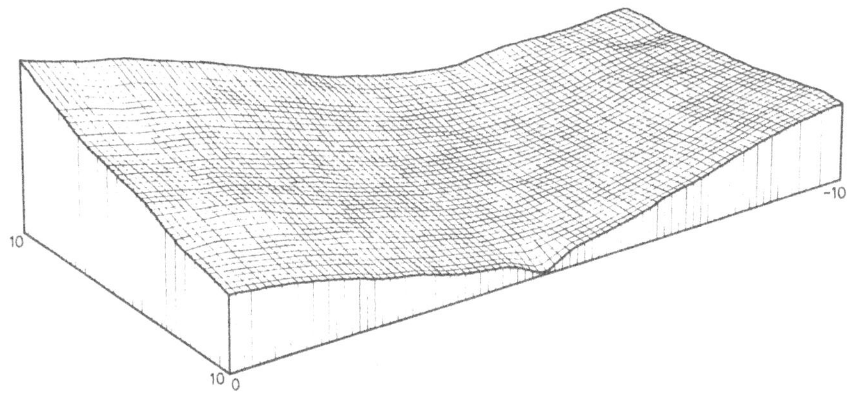


**Figure 8.** Perspective representation of sample semi-variogram of stone content at Plas Gogerddan viewed from a position above and to the left in Figure 7.

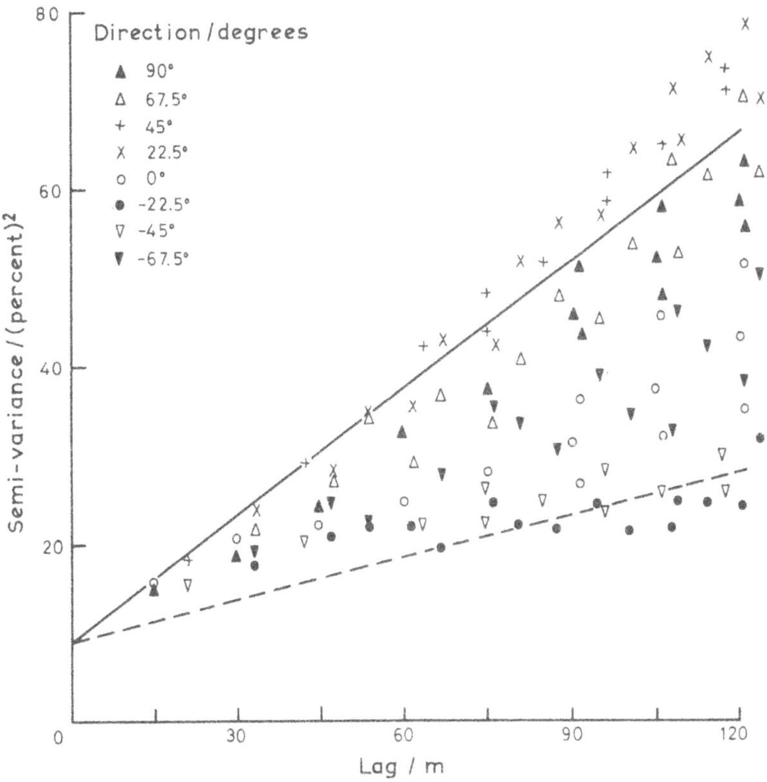


**Figure 9.** Cylindrical projection of two-dimensional sample semi-variogram of stone content with separate symbols for eight directions. The oblique lines form the envelope of the fitted linear model (see text).

The semi-variogram for one support can be related, at least theoretically, to that for another. In particular, there is often a practical interest in regularized semi-variograms. Suppose that $\gamma(\mathbf{h})$ is the semi-variogram for the support on which measurements were made. We shall consider this to be the punctual semi-variogram. Suppose also that we wish to determine a regularized semi-variogram, $\gamma_B(\mathbf{h})$ for larger blocks of size $B$. It can be shown that for a given lag $\mathbf{h}$,

$$\gamma_B(\mathbf{h}) = \bar{\gamma}(B, B_{\mathbf{h}}) - \bar{\gamma}(B, B), \quad [17]$$

where $\bar{\gamma}(B, B_{\mathbf{h}})$ denotes the average semi-variance between two blocks of size $B$ separated by the vector $\mathbf{h}$, and $\bar{\gamma}(B, B)$ is the average semi-variance within a block $B$; i.e., the within-block variance. If the distance $\mathbf{h}$ is much larger than the distance across the block $B$, then $\bar{\gamma}(B, B_{\mathbf{h}})$ approximately equals $\gamma(\mathbf{h})$, the point semi-variance. From equation [17] we obtain the useful approximation

$$\gamma_\beta(\mathbf{h}) \simeq \gamma(\mathbf{h}) - \bar{\gamma}(B, B). \quad [18]$$

Thus for $|\mathbf{h}| >> \sqrt{B}$ the regularized semi-variogram is derived from the punctual semi-variogram simply by subtracting the within-block variance.

This is especially pertinent for bulking. If the semi-variogram of a soil property measured on separate cores is known, then that for samples bulked over larger areas, the regularized semi-variogram, can be determined readily from it.

## IV. Semi-Variogram Models

Soil varies continuously in space, at least at most practical scales, and so semi-variograms of soil properties are continuous functions. The sample semi-variograms, however, consist simply of ordered sets of discrete values. These are estimates and as such are subject to error. They can be joined by straight lines or curves to give intermediate values, but the result is inevitably irregular. Nevertheless, a quick look at most well-estimated semi-variograms of soil will show that generally they approximate simple forms, and that it should be possible to fit simple functions to them. Later we shall wish to use the semi-variogram for estimation, and so its form is of both practical and scientific interest.

Figure 10, for one dimension, shows the principal features of semi-variograms of soil. In most instances it is found that $\gamma(h)$ increases from the smallest measured lag. In Figures 6 to 9 this increase in $\gamma(h)$ for stone content appears to have no limit: The soil appears to have no finite variance. This is represented by the solid line in Figure 10a. More often the semi-variance reaches a maximum at which it levels out, and such semi-variograms are said to be *transitive*. This maximum is known as the *sill*,

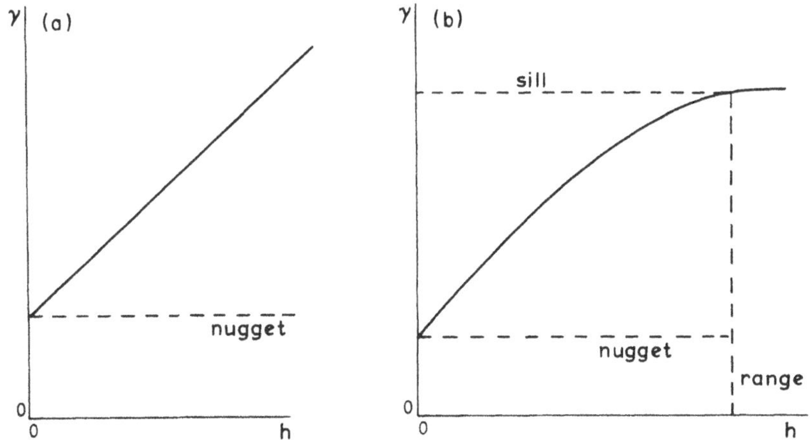


**Figure 10.** Elements of one-dimensional semi-variogram.

and its value is the *a priori* variance of the variable. The lag at which the sill is reached is known as the *range.* If $\gamma(h)$ approaches its maximum asymptotically, then for practical purposes the range may be chosen arbitrarily where $\gamma(h)$ is sufficiently close to its apparent sill.

The range is of considerable importance. It marks the limit of spatial dependence. The soil at places closer together than the range is related; at places further apart it is not, unless there is periodicity. The range may be interpreted as indicating the average distance across distinct soil types in some instances (Webster, 1973; Webster and Cuanalo, 1975; McBratney and Webster, 1981a). Experience to date suggests that at the common scales for detailed soil survey the range will usually be a few hundred meters. It does, however, depend on the size of the area sampled, and if the scale of the survey is changed substantially then the range is likely to do so too. The range also represents the limiting distance within which interpolation is worth attempting, and this will be discussed later.

By definition the semi-variance at zero lag is itself zero. But as can be seen from Figures 6, 12, 13, and 15, smooth curves approximating the sample semi-variances in these examples are unlikely to pass through the origin. All appear instead to approach positive finite intercepts on the ordinate at $h = 0$. This intercept is known as the *nugget variance*, and the phenomenon, which is widely recognized, is known as the *nugget effect*. The terms derive from gold mining. In gold-bearing rocks and sediments nuggets are sparse and small. Their diameters are very much smaller, by at least one order of magnitude, than the spacing between sampling cores. Most cores contain none; a small proportion contain one each, rarely more. Thus, the inclusion of a gold nugget in a drill core is regarded as a purely random event.

Few soil properties of common interest are quite like this, yet in practice their semi-variograms usually have distinct nugget variances. Indeed, in some investigations all the variation appears as nugget even when sampling as closely as 10 or 20 m. Campbell's (1978) results for the particle size fractions on both loess and till in Kansas were of this kind, as were several properties measured by McBratney and Webster (1981a) in northeast Scotland, e.g., Figure 12. Such results almost invariably mean that there is a source of spatial variation with a range much smaller than the smallest sampling interval. The shape of the semi-variogram in this range can be resolved only by much denser sampling.

### A. Positive Definite Functions

Thus we see that in choosing a model to represent a sample semi-variogram we must allow for at least three elements in most instances: an intercept, an increasing section of potentially varying shape, and a sill. In two dimensions there must also be provision for anisotropy.

Not any model that appears to fit the observed values will serve, however, for the following reason. Suppose that $Z(\mathbf{x})$ is a second-order stationary random function giving rise to the regionalized variable $Z(\mathbf{x}_i)$, $i = 1, 2, \ldots n$, and that its covariance function is $C(\mathbf{h})$ and $\gamma(\mathbf{h})$ its semi-variogram. And consider the linear combination

$$Y = \sum_{i=1}^{n} \lambda_i Z(\mathbf{x}_i), \qquad [19]$$

where $\lambda_i$, $i = 1, 2, \ldots, n$, are any arbitrary weights. This quantity is a random variable with

$$\text{var}[Y] = \sum_{i=1}^{n} \sum_{j=1}^{n} \lambda_i \lambda_j C(\mathbf{x}_i, \mathbf{x}_j). \qquad [20]$$

The variance of $Y$ may be positive or zero, but it may not be negative, and the covariance function on the right-hand side of equation [20] must ensure that this condition is met. The covariance, $C(\mathbf{h})$, must be a positive definite function, and only functions that meet this criterion are acceptable.

As we have seen, there are situations where soil properties do not have definable covariances because their variances increase apparently without limit. Provided the intrinsic hypothesis holds, however, we can make use of the following relation. Equation [20] can be rewritten as

$$\text{var}[Y] = C(0) \sum_{i=1}^{n} \lambda_i \sum_{j=1}^{n} \lambda_j - \sum_{i=1}^{n} \sum_{j=1}^{n} \lambda_i \lambda_j \gamma(x_i, \mathbf{x}_j). \qquad [21]$$

The first term on the right-hand side of this equation can be eliminated provided

$$\sum_{i=1}^{n} \lambda_i = 0, \qquad [22]$$

giving

$$\text{var}\,[Y] = -\sum_{i=1}^{n}\sum_{j=1}^{n} \lambda_i\lambda_j\,\gamma(\mathbf{x}_i, \mathbf{x}_j). \qquad [23]$$

This too must be non-negative, and we are left with the constraint on the semi-variogram that it must be a positive definite function with the added condition that the weights in equation [23] sum to 0.

Although we may state the above, it is not easy to test whether any particular model is conditional positive definite, and it can be difficult to create "data" that will demonstrate that a model is inappropriate (Armstrong and Jabin, 1981). The standard approach involves examining the Fourier transform of the semi-variogram or covariogram, and Christakos (1984) has listed the conditions that the spectrum must meet for a model to be acceptable. Dunn (1983) has shown that a simpler test can be applied if a model is required only for lags at which there are observed data. This, however, will be too restrictive in many instances.

The consequences of fitting a function that is not positive definite are not always disastrous. The investigator does not necessarily encounter negative variances, and so remains unaware of the dangers. The need to avoid such models is not always appreciated, especially because a function that is positive definite in one dimension is not necessarily so in two or three dimensions.

## B. Safe Models

This section defines models that can be recommended for semi-variograms of soil properties. They are defined for one dimension but are safe in the sense that they are conditional positive definite in two and three dimensions. In the French literature they are often referred to as *authorized models*.

### 1. Linear Models

The simplest model that can be fitted in one dimension is clearly linear. It has slope $w$ and may have an intercept or nugget variance $c_0$. Its formula is

$$\begin{aligned} \gamma(h) &= c_0 + wh \qquad \text{for } h > 0 \\ \gamma(0) &= 0. \end{aligned} \qquad [24]$$

Notice that it has no sill. Figure 11 shows an example. A linear model is also fitted to the first 15 sample semi-variances of pH in Figure 13.

In the limit $w$ can be zero (Figure 12). The semi-variogram is then said to show a pure nugget effect. There is no spatial dependence at the scale of investigation since all of the variance occurs within the smallest sampling interval.

2. Spherical Models

A model that has been found to fit not only many semi-variograms of soil properties but also those of mineral deposits of many kinds is the spherical model. Its definition is

$$\gamma(h) = c_0 + c\left[\frac{3}{2}\frac{h}{a} - \frac{1}{2}\left(\frac{h}{a}\right)^3\right] \qquad \text{for } 0 < h \leq a$$

$$\gamma(h) = c_0 + c \qquad \text{for } h > a$$

$$\gamma(0) = 0. \qquad [25]$$

Its characteristics are illustrated in Figure 13, in which $a$ is the range, $c_0$ +

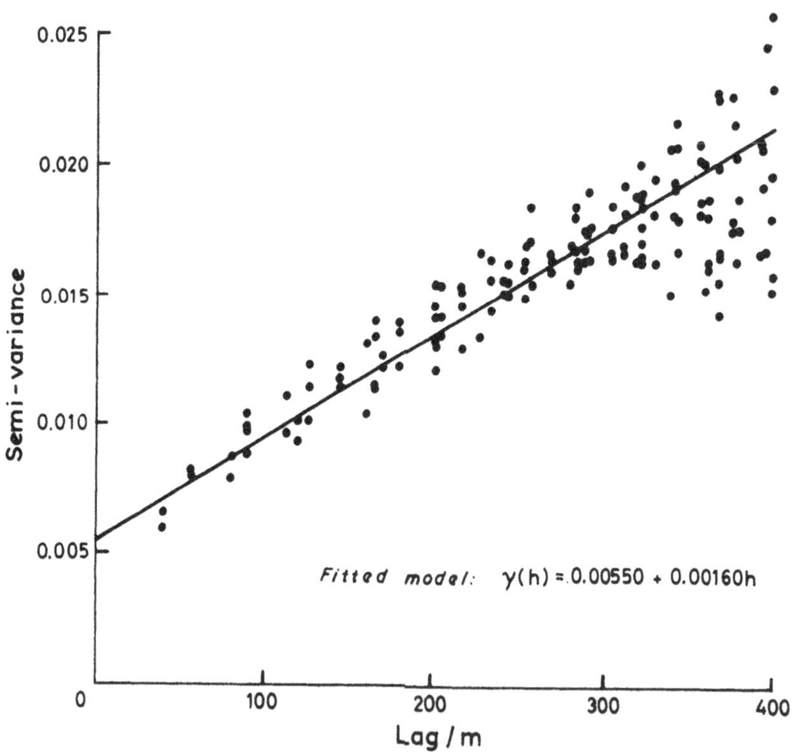


**Figure 11.** Semi-variogram of exchangeable potassium in topsoil, measured in μg/g and transformed to common logarithms, at Broom's Barn with linear model fitted (see Webster, 1981).

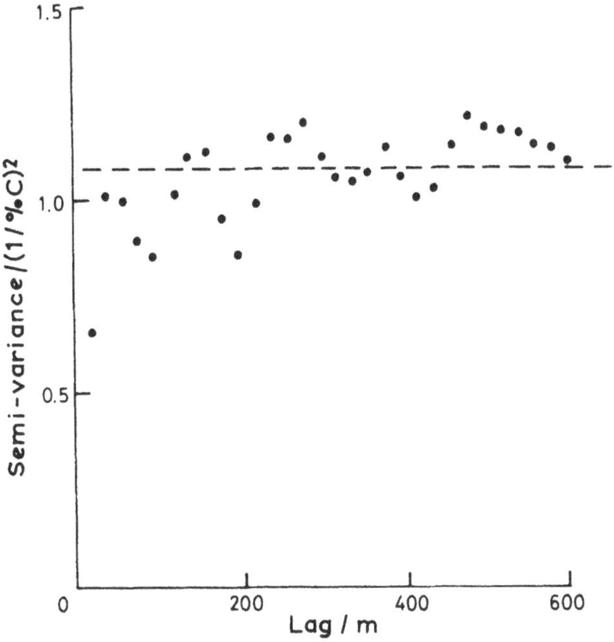


**Figure 12.** Semi-variogram of organic carbon in the subsoil, measured as percentage and transformed to reciprocal, at Tillycorthie, Aberdeenshire (see McBratney and Webster, 1981a).

$c$ is the sill, and $c_0$ is the nugget variance as before. Its tangent at $h = 0$ cuts the sill at ⅔ $a$. Figure 13 shows the sample semi-variogram of pH over 420 m well fitted by the spherical model.

Theoretically, the spherical model derives from a moving average random process. It is the three-dimensional form of a quite general $n$-dimensional model (Matérn, 1960). The idea is that in three dimensions a random function, of which the property measured is a realization, depends on the volume of the intersection of two spheres of equal radius. If $a$ is the diameter of the spheres and $h$ the distance between their centers then this volume, $V$, is

$$V = \frac{\pi}{4}\left(\frac{2a^3}{3} - a^2h + \frac{h^3}{3}\right) \qquad \text{for } h \leq a. \qquad [26]$$

If this is expressed as a fraction of the volume of a sphere by dividing by $\frac{1}{6}\pi a^3$ we obtain the autocorrelation function,

$$\rho(h) = 1 - \frac{3}{2}\frac{h}{a} + \frac{1}{2}\frac{h^3}{a^3}, \qquad [27]$$

from which the semi-variogram is derived using the relation defined in equation [11].

Such a function would seem fairly obviously applicable to three-

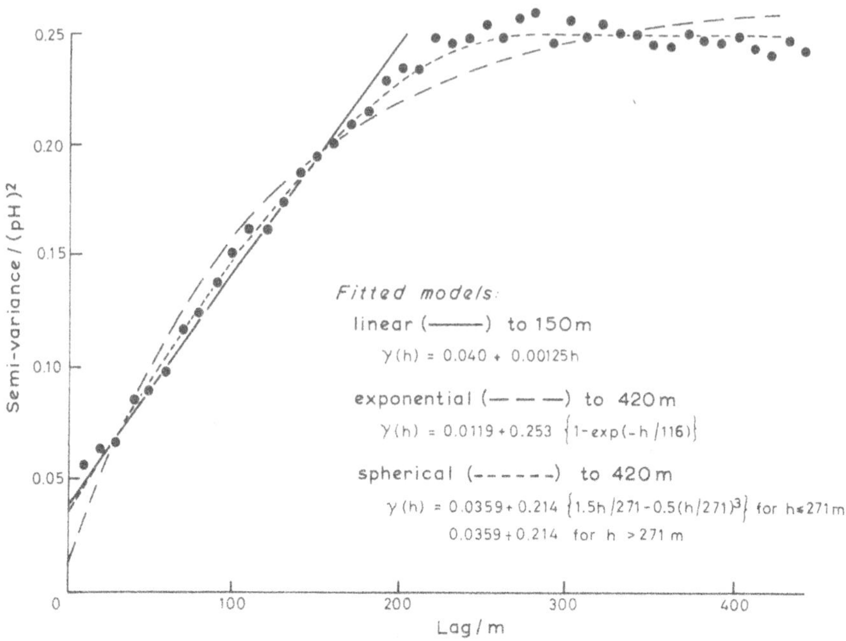


**Figure 13.** Sample semi-variogram of pH in the topsoil at Sandford St. Martin with linear, exponential, and spherical models fitted by least-squares approximation.

dimensional rock bodies, and it has indeed proved so. It is less obviously relevant for distributions in one and two dimensions, and yet it nearly always fits experimental results from soil sampling better than the one- and two-dimensional analogs, e.g., Figure 14, which are described later. The reason is presumably that there are additional sources of variation in soil at other spatial scales.

Similar interpretations of variograms near Witney (Webster, 1973) and Sandford (Webster and Cuanalo, 1975), both in Oxfordshire, and of semi-variograms at Tillycorthie in northeast Scotland (McBratney and Webster, 1981a) led the authors to seek distinct parcels of different soil type with linear dimensions equal to the range. In all instances they were able to find boundaries at approximately this average spacing using statistical search procedures devised by Webster (1978) and Hawkins and Merriam (1974), and so confirm their interpretations.

## 3. Exponential Models

The formula of the exponential model is

$$\begin{aligned} \gamma(h) &= c_0 + c\,[1 - \exp(-h/r)] \qquad \text{for } h > 0 \\ \gamma(0) &= 0. \end{aligned} \qquad [28]$$

The spatially dependent variance and nugget are $c$ and $c_0$ as before, and $r$ is a distance parameter controlling the spatial extent of the function. Here $\gamma(h)$ approaches the sill asymptotically, and so there is no strict finite range. Nevertheless, it is clear that for practical purposes the semi-variance ceases to increase beyond some point, and a commonly used rule of thumb is to take this as the effective range, $a' = 3r$ at which $\gamma(a')$ is then equal to approximately $c_0 + 0.95c$. Figure 13 shows the form of the model. The curve is the best fitting exponential to the sample semi-variogram of pH but clearly does not represent the sample values well.

The exponential model has an important place in statistical theory. It represents the essence of randomness in space. It is the semi-variogram of first-order auto-regressive and Markov processes. As autocorrelation function it has been the basis of several theoretical studies of the efficiency of sampling designs, by for example, Cochran (1946), Yates (1948), Quenouille (1949), and Matérn (1960). Semi-variograms of exponential form are also to be expected where differences in soil type are the main contributors to soil variation and where soil boundaries occur as a Poisson process. Burgess and Webster (1984) and Webster and Burgess (1984a) found the latter to be so along transects in many instances. If the intensity of the process is $\alpha$, then the mean distance between the boundaries is $\bar{d} = 1/\alpha$, and the semi-variogram is

$$\begin{aligned}\gamma(h) &= c\,[1 - \exp(-h/\bar{d})]\\ &= c\,[1 - \exp(-\alpha h)].\end{aligned} \qquad [29]$$

Oliver (1984), working in the Wyre Forest of England, obtained exponential semi-variograms for a number of soil properties all with approximately the same distance parameter. She attributed their form largely to this process.

Sisson and Wierenga (1981) found that the rate at which water filtered into soil varied exponentially with distance and postulated it as a first-order auto-regressive process. On a very different scale Yost *et al.* (1982) estimated the semi-variograms of several chemical properties to 60 km over the Island of Hawaii and were able to fit exponential models in all cases. Similarly Xu and Webster (1984) found the pH of the soil in Zhangwu, a region covering 3500 km$^2$ in northeast China, to be distributed exponentially.

4. The Hyperbola

A novel function in this group is that used by Vieira *et al.* (1981) to describe variation in infiltration across irrigated land in the Central Valley of California, and shows the strong influence of physical chemistry in soil science even in this field. The function is a hyperbola,

$$\gamma(h) = \frac{h}{\alpha + \beta h}, \qquad [30]$$

which is the equation of the Langmuir adsorption of isotherm, with $\gamma(h)$ replacing the mass adsorbed and $h$ the pressure. The model is transitive: It approaches the sill value $1/\beta$ asymptotically and has a limiting gradient of $1/\alpha$ at the origin. Nielsen[1] has stated that it is positive definite in two dimensions. For infiltration rate in mm $hr^{-1}$ and lag in m, Vieira *et al.* found values for $\alpha$ and $\beta$ as 1.207 and 0.105, respectively. There was no intercept, though it would be perfectly in order to add a nugget term.

## C. Risky Models

In most instances soil properties appear transitive: The semi-variogram appears monotonic increasing to a sill, and this is to be expected in a finite region. The main difference among them is the degree of curvature. The exponential function curves gradually. The spherical model curves more tightly. But there are instances where the semi-variogram appears to curve more tightly still, even abruptly, and the investigator may be tempted to fit more tightly curving models. We consider two below.

### 1. The Circular Model

Just as we obtained the spherical model for three dimensions we can derive a circular model in two (Zubrzycki, 1957; Dalenius *et al.*, 1961). The area of intersection of two equal circles is given by

$$A = \frac{a^2}{2}\cos^{-1}\left(\frac{h}{a}\right) - \frac{h}{2\pi}\sqrt{a^2 - h^2} \qquad \text{for } h \leq a, \qquad [31]$$

where $a$ is the diameter of the circles and $h$ the distance between their centers. Expressing this as a fraction of the area of the circle gives

$$\rho(h) = \frac{2}{\pi}\left[\cos^{-1}\left(\frac{h}{a}\right) - \frac{h}{a}\sqrt{1 - \frac{h^2}{a^2}}\right] \qquad [32]$$

for the autocorrelation function, and the following semi-variogram:

$$\begin{aligned} \gamma(h) &= c_0 + c\left[1 - \frac{2}{\pi}\cos^{-1}\left(\frac{h}{a}\right) + \frac{2h}{\pi a}\sqrt{1 - h^2/a^2}\right] && \text{for } 0 < h \leq a \\ \gamma(h) &= c_0 + c && \text{for } h > a \\ \gamma(0) &= 0. && [33] \end{aligned}$$

[1]Personal communication.

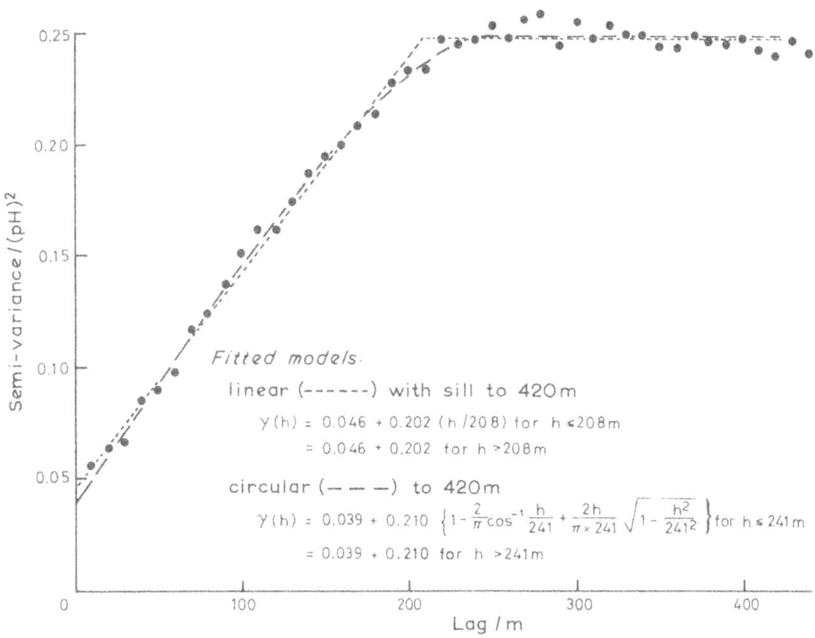


**Figure 14.** Sample semi-variogram of pH in the topsoil at Sandford St. Martin. Linear-with-sill and circular models are fitted by least-squares approximation.

This model and its consequences for plane sampling were explored in detail by Dalenius *et al.* (1961). Figure 14 shows it fitted to the sample semi-variogram of sand content at Sandford. It clearly describes the results well.

## 2. Linear Model with Sill

The extreme form of transitive model is linear with a sill:

$$\begin{aligned} \gamma(h) &= c_0 + c\left(\frac{h}{a}\right) && \text{for } 0 < h \leq a \\ \gamma(h) &= c_0 + c && \text{for } h > a \\ \gamma(0) &= 0. && \end{aligned} \qquad [34]$$

I have expressed it in this way to emphasize its similarity to that of the spherical and circular models. Again, by analogy with the spherical scheme this can arise from linear zones of influence of equal length but with varying distance between their centers, $h$. For a given length, $a$, the overlap is simply $a-h$, or as a proportion $1-h/a$, provided $h \leq a$.

There are several examples in the literature where investigators have used this form of model for semi-variograms of soil. Figure 14 shows it fitted to the same Sandford semi-variances as above, and again the fit is reasonable.

What, then, are the risks of using these models? They concern the positive definite condition. The circular model is positive definite in one and two dimensions, but not in three. This will usually not matter in soil survey, which is essentially concerned with variation in the plane. The linear model with sill is positive definite in one dimension, but not in more. This model should not be used in soil survey, however well it might appear to fit, unless variation is strictly limited to just the one dimension. Those investigators who have used the model in two-dimensional contexts have either not examined the variances of combinations of values or have had the fortune not to encounter negative variances in practice.

### D. Nested Models

This paper began with a study of nested sampling and an analysis based on a nested model with variation deriving from several sources with different spatial scales. The components of variance measured the amount of variance contributed by each scale, and by accumulating them we were able to show how variance increased with increasing distance. Starting at the bottom of the hierarchy, the variance at the spacing of level 3 consists of the sum of components $\sigma_4^2$ and $\sigma_3^2$, that at level 2 consists of $\sigma_4^2 + \sigma_3^2 + \sigma_2^2$, and so on. Miesch (1975) showed that this is precisely the same as in the semi-variogram: The variance at a lag of 2, say $\gamma(2)$, comprises that at lag 1, $\gamma(1)$ or $\sigma_1^2$, plus a component deriving from distances between 1 and 2, say, $\sigma_2^2$. We have, therefore,

$$\gamma(2) = \gamma(1) + \sigma_2^2. \qquad [35]$$

Thus, the cumulative variances from the nested analysis constitute a semi-variogram.

Most of the semi-variograms that we have seen so far are already nested models in the sense that they have nugget variances. They can be represented by

$$\tfrac{1}{2}\,\mathrm{var}\,[z(x) - z(x + h)] = \gamma(h) = \gamma_0(h) + \gamma_1(h). \qquad [36]$$

In other words, the semi-variogram is the sum of two functions, one representing pure nugget, $\gamma_0(h)$, and the other a spatially dependent one, $\gamma_1(h)$. This can be extended by adding further functions.

In studies of mineralization it is often found that one more term is needed to describe the variation adequately; thus,

$$\gamma(h) = \gamma_0(h) + \gamma_1(h) + \gamma_2(h). \qquad [37]$$

Mine surveyors and geochemists have found the double spherical nested model especially valuable, and my colleagues and I have fitted it to several sample semi-variograms of soil properties (McBratney *et al.*, 1982;

Webster and Nortcliff, 1984). Figure 1.18 is an example. In these $\gamma_1(h)$ and $\gamma_2(h)$ are separate semi-variograms:

$$\gamma_1(h) = c_1\left[\frac{3}{2}\frac{h}{a_1} - \frac{1}{2}\left(\frac{h}{a_1}\right)^3\right] \quad \text{for } 0 < h \leq a_1$$
$$\gamma_1(h) = c_1 \quad \text{for } h > a$$
$$\gamma_2(h) = c_2\left[\frac{3}{2}\frac{h}{a_2} - \frac{1}{2}\left(\frac{h}{a_2}\right)^3\right] \quad \text{for } 0 < h \leq a_2$$
$$\gamma_2(h) = c_2 \quad \text{for } h > a_2. \qquad [38]$$

The two sills $c_1$ and $c_2$ are in general different. The ranges will usually be very different; otherwise, they are unlikely to be distinguished.

The quantity $\gamma_0(h)$ is the nugget variance: a semi-variogram with its effective range much less than the smallest sample interval measured.

These three semi-variograms comprise the nested model fitted to the semi-variogram of readily extractable copper content in the topsoil of southeast Scotland and are shown separately in Figure 15, from McBratney *et al.* (1982). The coefficients are given below. The variances are in units of ($\log_{10}$ μg Cu/g soil)$^2$ and the ranges in km.

| | |
|---|---|
| $c_0$ (nugget) | 0.0213 |
| $c_1$ | 0.0257 |
| $c_2$ | 0.0196 |
| $a_1$ | 2.26 |
| $a_2$ | 15.5 |

Recently Burrough (1983b) has postulated a nested structure for soil variation that arises from independent soil-forming factors that have operated over distinctly different spatial scales. He mentions the effects of geology, relief, and earthworms, to which one can add those of tree-throw and manmade divisions into fields and farms as examples of other scales. The effects can be ranked according to their ranges, and Burrough proposed a geometric scale in much the same way as Youden and Mehlich (1937) did for their nested sampling. As an example Burrough chose ranges, $a$, on a scale such that $a_1 = \frac{1}{2}a_2 = \frac{1}{4}a_3 = \frac{1}{8}a_4 \ldots$ Thus in one dimension the semi-variogram has the general form of equation [37] comprised as

$$\gamma(h) = c_1\frac{h}{a_1} + c_2\frac{h}{a_2} + c_3\frac{h}{a_3} + c_4\frac{h}{a_4} + \ldots. \qquad [39]$$

In other words, it consists of a series of linear semi-variograms with sills that together form a linear spline. Burrough estimates the sill values, $c_i$, by solving a system of linear equations and shows how such models fit actual one-dimensional semi-variograms. The fits are undoubtedly good. His

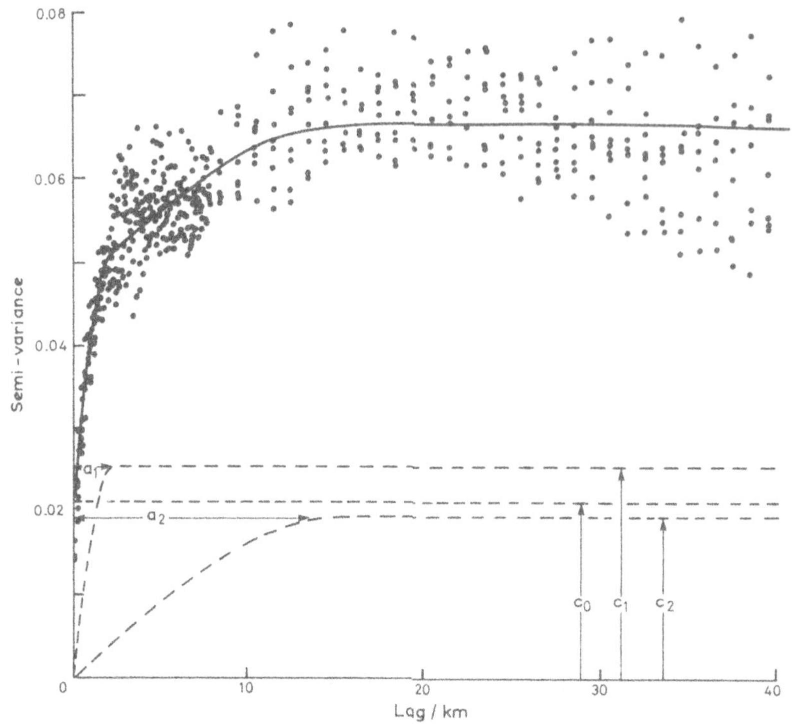


**Figure 15.** Semi-variogram of available copper in the topsoil of southeast Scotland. The three component semi-variograms of the nested spherical model are shown by the dashed lines (see McBratney *et al.*, 1982).

expression for the semi-variogram of pH in the topsoil at Sandford, Figure 14, converted into the terms of equation [34] is

$$\begin{aligned} \gamma(h) &= 0.048 + 0.200\frac{h}{209} && \text{for } 0 < h < 209 \\ \gamma(h) &= 0.048 + 0.200 && \text{for } h > 209 \\ \gamma(0) &= 0, && \end{aligned} \qquad [40]$$

which agrees very closely with the model I fitted in Figure 14.

The model is postulated specifically for one dimension. In more dimensions linear splines are not generally positive definite, and the component semi-variograms should be replaced by their two- or three-dimensional analogs, the circular or spherical models.

## E. Models for Anisotropy

As discussed earlier, soil does not necessarily vary equally in all lateral directions. Figures 6 to 9 showed an example in which the stone content varied anisotropically. Another example, Figure 21, appears later, and McBratney and Webster (1983) analyzed a further case, Figure 17.

The semi-variogram for the individual directions in Figure 6 are very close to linear, though with different slopes. The contours of the semi-variogram, Figure 7, are approximately elliptical, and this suggests that the semi-variogram can be represented by a simple sinusoidal function. Burgess and Webster (1980a) initially represented this semi-variogram using polar coordinates by

$$\gamma(h, \theta) = c_0 + u(\theta)|\mathbf{h}|, \qquad [41]$$

where

$$u(\theta) = A \cos^2(\theta - \phi) + B \sin^2(\theta - \phi). \qquad [42]$$

The parameters are $\phi$, the direction of maximum variation; $A$, the gradient of the semi-variogram in that direction; and $B$, the gradient in the direction $\phi + \pi/2$.

Burgess *et al.* (1981) later modified this to

$$u(\theta) = [A^2 \cos^2(\theta - \phi) + B^2 \sin^2(\theta - \phi)]^{1/2}. \qquad [43]$$

This last expression now defines strict geometric anisotropy (David, 1977), which can be made isotropic by a linear transformation of the coordinate system. It might help to appreciate this by imagining the land on a rubber sheet. By stretching the sheet in direction $\phi$ in the proportion $A$ to $B$ the semi-variogram in that direction will have the same gradient as that in the perpendicular direction. The proportion $A/B$ can be regarded as the anisotropy ratio.

On fitting this model to the sample values, Burgess *et al.* (1981) obtained the following:

$$\gamma(h, \theta) = 9.06 + [7.18^2 \cos^2(\theta - 1.00) + 2.40^2 \sin^2(\theta - 1.00)]^{1/2}|\mathbf{h}|. \qquad [44]$$

As above, this would appear simply as a series of concentric elliptical isarithms in plan. On the cylindrical projection it can be represented as a defining envelope, shown by the lines of maximum and minimum gradient in Figure 9. These are

$$\gamma_1(h) = 9.06 + 7.18\,h$$

and

$$\gamma_2(h) = 9.06 + 2.40\,h, \qquad [45]$$

respectively. Where the anistropy is geometric, this form of display contains all the information in the semi-variogram except the orientation, and that can be stated.

In the linear case the quantity $u(\theta)$ is the gradient in direction $\theta$. In transitive models the gradient $u(\theta)$ in equation [42] is replaced by a range parameter, say $r(\theta)$, and $A$ and $B$ then represent the maximum and minimum ranges of the model. No one seems to have encountered semi-variograms of this kind so far for soil.

A further type of anisotropy of which soil scientists must be aware is where the property of interest has different sills in the different directions. This is known as zonal anisotropy (Journel and Huijbregts, 1978). Again there are no reports of its occurrence in the horizontal plane in the soil literature. Almost certainly it occurs when the vertical dimension is added, but no one has analyzed three-dimensional soil bodies or even vertical planes in this way yet to my knowledge.

## V. Fitting Models

Choosing models to describe observed semi-variances and the procedures for fitting them are matters of some controversy. Indeed, they are more controversial than any other aspect of spatial analysis covered in this review.

The choice of model will obviously be governed by the general graphic appearance of the sample semi-variogram. For example, one will choose a function for pH at Sandford that turns more tightly than the exponential curve, Figure 13. But not any function that appears to fit will do: As above, functions must be positive definite.

An investigator will also be influenced by what is already known about the soil of a region—by the form that the semi-variogram is expected to take. The investigator may wish to interpret the semi-variogram in terms of some physical process or geological origin, and a model that makes this possible is likely to be preferred to one that does not.

The choice of model will also depend on the purpose for which it is wanted. The criteria for choosing models for estimation or interpolation might be quite different from those used for illustration or explanation. An investigator may also fit different models for different-sized regions depending on the maximum lag that is of interest (Webster and Burgess, 1984b).

Having decided the form of the model and the lag over which it is to be fitted, there is then a choice of fitting procedure. The linear model is, of course, easy to fit by simple regression. Usually the number of pairs of comparisons in the estimated semi-variances differ from one another, and so the regression should be weighted accordingly. Some models can be made linear by transformation and the parameters estimated again by regression. The Langmuir isotherm, for example, in the form given in equation [30] can be transformed to

$$\frac{h}{\gamma} = \alpha + \beta h, \qquad [46]$$

from which $\alpha$ and $\beta$ are readily found.

Most of the useful models are non-linear, and so a less straightforward approach must be adopted. Journel and Huijbregts (1978) recommend procedures for finding the nugget variance and sill of transitive models. Many sample semi-variograms are approximately linear over the first few lags, and so by fitting straight lines and extrapolating intercepts on the ordinate can be found. As Figure 13 shows, this value is much the same as that obtained by non-linear least-squares approximation of the whole curve.

Their recommendation for finding a sill is less sound, mainly because of misunderstanding among readers. It will pay, therefore, to make a digression at this point to clarify matters.

The basic notion of transitive variation is that there is a distinct range within which variation increases with increasing lag distance, $h$. Beyond this range $\gamma(h)$ remains constant, however much larger the lag. In principle the semi-variance remains constant to infinity at what is known as the *a priori* variance of the property of interest. One aim in sampling and model fitting is to estimate this *a priori* variance by the sill of the semi-variogram.

If we were able to sample in infinite space then the spatial dependence that occurs within limited lags would be negligible. We could disregard it. The *a priori* variance would be equivalent to the population variance, $\sigma^2$, of classical statistics, which we could estimate in the usual way by

$$s^2 = \frac{1}{n-1}\sum_{i=1}^{n} [z(\mathbf{x}_i) - \bar{z}]^2, \qquad [47]$$

where $\bar{z}$ is the sample mean. Of course, we never have an infinite space in which to sample, but if the region, $B$, is very large in relation to the range of the semi-variogram, the regional variance, say $\sigma_B^2$, and our estimate of it, $\hat{\sigma}_B^2$, might be so close to the *a priori* variance that it will serve.

In practice, however, we are more often concerned with situations in which the range of the semi-variogram is a considerable proportion of the distance across the region. Now the population variance of a finite region, or *dispersion variance* as it is known to geostatisticians, is the average semi-variance within that region. Formally it is

$$\sigma_B^2 = \frac{1}{B^2}\int_B \int_B (\mathbf{x} - \mathbf{x}')\, d\mathbf{x}\, d\mathbf{x}', \qquad [48]$$

where $\mathbf{x}$ and $\mathbf{x}'$ are two points that describe the region independently. It is represented in Figure 16 by the area under the curve. Clearly, this is less than the area beneath the sill. In fact the dispersion variance, which one would estimate by equation [47] from a sample is that value which makes the two hachured areas equal. It must be less than the sill, unless the semi-variogram is pure nugget at the working scale. Thus, the practice of

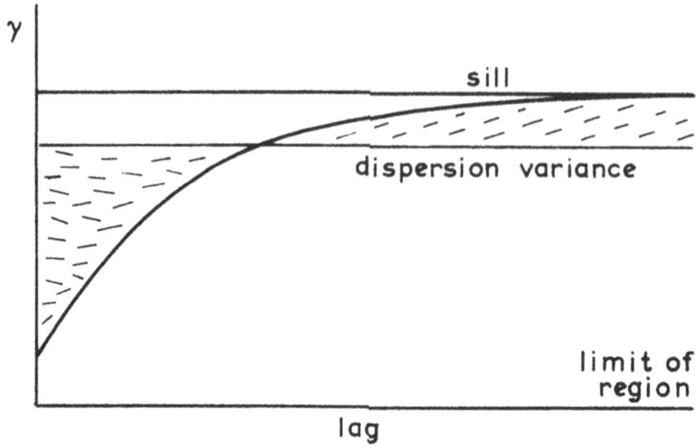


**Figure 16.** Transitive semi-variogram and dispersion variance of a finite region.

estimating the sill by the sample variance within a region is not to be recommended.

### A. Recommended Practice

Despite the criticisms of automatic least-squares fitting made by Journel and Huijbregts (1978) and others, my colleagues and I have found that it works well where there are sufficient data. For non-linear models we first obtain values for the non-linear parameters, notably the range, iteratively and then fit the remainder by solving a matrix equation. We weight the semi-variances by the number of pairs of comparisons used to estimate them. We use the Maximum Likelihood Program, MLP, written by Ross (1980) at Rothamsted for this and program our own models into it. Other sound suggestions are made by Armstrong (1984) in a recent discussion of the problem. She clearly favors a weighted area method involving Laplace transforms that is especially well suited for exponential semi-variograms. Kitanidis (1983) discusses the problem further and proposes a true maximum likelihood solution for fitting parameters to the whole set of covariances among the data. It is likely to be restricted to fairly small sets of observations, however. Readers wishing to analyze spatially distributed soil data should read these papers and also that by Delfiner (1976).

## VI. Fractal Representation

We have already remarked on the nested nature of soil variation. We noted that soil can apparently vary increasingly without limit as the area increases. At the other end of the scale most semi-variograms recorded on point supports have nugget variances. There always seems to be some variance within the smallest sampling interval, and more to encounter if we go further afield. It is clear, therefore, that the semi-variogram obtained in any study depends very much on the scale of the study. There is no absolute semi-variogram for a soil property.

Gajem *et al.* (1981) illustrate this dependence on scale well. They examined a region of 85 ha at three scales by sampling on transects at 20 cm, 2 m, and 20 m intervals and increasing the lengths of the transects in proportion. As the length of transect increased so did the variance and the range of dependence.

Such facts led Burrough (1983a) to consider whether soil exhibited true similarity at different scales and to explore its representation by fractals (Mandelbrot, 1982). He took as starting point the stochastic fractal of linear Brownian motion,

$$z(x) = z(x + h) + \varepsilon, \qquad [49]$$

where $\varepsilon$ is a Gaussian random deviate. Its semi-variogram has the one-dimensional form,

$$\mathrm{E}\{[z(x) - z(x + h)]^2\} = 2\gamma(h) = h^{2H}, \qquad [50]$$

where $H = 0.5$. If the sampling interval is divided by any arbitrary positive value $r$ and the result rescaled in the ratio $r^H$ then the new semi-variogram will be identical with the old one. In this sense Brownian motion is a self-similar or fractal process.

In ordinary Brownian motion the successive values of $\varepsilon$ are totally independent. Mandelbrot, however, showed that by increasing $H$ in the range 0.5 to 1 he could generate a family of trails smoother than Brownian motion in which successive values of $\varepsilon$ were positively correlated. Conversely, if he diminished $H$ between 0.5 and 0 so that the $\varepsilon$ were negatively correlated, then the tracks were noisier than those of Brownian motion.

At the limit $H = 1$ the semi-variogram is a parabola and variation is smooth and differentiable. At the other extreme $H = 0$ indicates pure noise, which seems impossible in continuous space. Indeed, Journel and Huijbregts (1978) preclude these limiting values: The function $\gamma(h) = -h^2$ is not conditional positive definite. Following Orey (1970) and Berry and Lewis (1980) Burrough points out that $H$ is related to the Hausdorff-Besicovitch or fractal dimension $D$ by $H = 2-D$, and by plotting the semi-variogram on double logarithmic graph paper he estimates $D$ from the slope, $m = 4 - 2D$.

Burrough estimated $D$ for a number of soil properties in several different regions and obtained values ranging from 1.5 to 2. His values of $H$ therefore lay between 0 and 0.5, suggesting that soil varied in a way more random than Brownian motion. While this may be so, Burrough concluded that the different independent factors that affect soil variation combine in a way that overshadow any self similarity. The fractal Brownian model seems less appropriate for soil than the nested models described above

(Burrough, 1983b). Nevertheless, the fractal representation should be borne in mind by soil scientists, and $D$ values of soil properties computed for reference against those of Brownian fractals (Burrough, in press).

## VII. Cross Correlation

So far we have considered the spatial correlation of a soil property with itself; i.e., its autocorrelation. The same concepts extend to two properties, so that the values of the one depend in the statistical sense on those of the other at other places nearby. Where this is so the variables are said to be co-regionalized or cross correlated.

If we adhere to the intrinsic hypothesis as the strongest assumption then we can express the spatial interdependence of two variables, say $u$ and $v$, by their cross semi-variogram, $\gamma_{uv}(\mathbf{h})$. The cross semi-variance is defined by

$$\gamma_{uv}(\mathbf{h}) = \mathrm{E}\{[z_u(\mathbf{x}) - z_u(\mathbf{x} + \mathbf{h})]\ [z_v(\mathbf{x}) - z_v(\mathbf{x} + \mathbf{h})]\}, \qquad [51]$$

and is estimated like the auto semi-variance by

$$\hat{\gamma}_{uv}(\mathbf{h}) = \frac{1}{2N(\mathbf{h})} \sum_{i-1}^{N(\mathbf{h})} [z_u(\mathbf{x}_i) - z_u(\mathbf{x}_i + \mathbf{h})]\ [z_v(\mathbf{x}_i) - z_v(\mathbf{x}_i + \mathbf{h})], \qquad [52]$$

where $N(\mathbf{h})$ is the number of pairs of comparisons that can be made at lag $\mathbf{h}$. Sample cross semi-variograms can be computed in one dimension from transect data and in two dimensions from grids, and the same devices can be adopted for dealing with missing data and less regular sampling as are used for single variables.

There is, however, one feature of cross correlation that does not occur with single variables. It concerns the symmetry of the relation. If we assume that the two properties are second-order stationary and consider their relations in one dimension, then we can define the cross covariance $C_{uv}(h)$ at lag $h$ as

$$C_{uv}(h)\ \mathrm{E}\{[z_u(x) - \bar{z}_u]\ [z_v\ (x + h) - \bar{z}_v]\}. \qquad [53]$$

The cross correlation is then

$$\rho_{uv}(h) = C_{uv}\ (h) /\ [C_{uu}(0) \cdot C_{vv}(0)]^2, \qquad [54]$$

and the cross semi-variance is related to it by

$$\begin{aligned} 2\gamma_{uv}(h) &= 2C_{uv}(0) - C_{uv}(h) - C_{vu}(h) \\ &= 2C_{uv}(0) - C_{uv}(-h) - C_{vu}(-h). \end{aligned} \qquad [55]$$

If we interchange $u$ and $v$ in either equation [51] or equation [55] for $\gamma_{uv}(h)$ it makes no difference. If, however, we interchange them in the equation for the cross covariance, equation [53], we obtain a different result in general.

The value of $C_{uv}$ $(h)$ is only incidentally the same as $C_{vu}$ $(h)$. Small differences in their estimates can be regarded as sampling effects, but any large difference means that one variable lags behind the other. In these circumstances the coregionalization can be represented fully only by the cross covariogram.

The same rules for fitting models apply to cross semi-variograms as to auto semi-variograms. The models must be such that any linear combination of values,

$$Y_k = \sum_{i=1}^{n} \lambda_{ki} \, z_k(\mathbf{x}_i), \qquad [56]$$

has a non-negative variance for $k = u$ and $k = v$, and indeed for any other variables in the model. This implies that at any lag $\mathbf{h}$ the cross covariance matrix of the variables is positive definite. The matter is somewhat complex, and readers who wish to delve more deeply should consult Journel and Huijbregts (1978).

Vauclin *et al.* (1983) and McBratney and Webster (1983) have analyzed co-regionalizations of soil properties, and the following example from the latter paper illustrates the techniques.

McBratney and Webster examined the cross correlation of particle size fractions in the topsoil and subsoil of a 10.6 ha field on the Woburn Experimental Station in central England. The field had been sampled at the nodes of a 10 m square grid, and cores 6 cm in diameter were taken from 0 to 20 cm and 40 to 60 cm depth. The percentages of sand (2 mm to 63 μm), silt (63 μm to 2 μm), and clay (< 2 μm) were determined on every subsoil core but on only every fourth sample of topsoil. The percentages were transformed to logit ($\ln[p/(100 - p)]$ where $p$ is a percentage) to stabilize variances.

As might be expected there was some strong simple correlation among the fractions. This was, however, restricted largely to the sand and silt, for which correlation coefficients were all larger than 0.8 in absolute value. There was also a marked anisotropy, and a part of the analysis was concerned to establish the extent to which this was common to all the particle size fractions.

All the semi-variograms were linear, at least to 100 m, and so had the form of equation [44]. Table 3 lists the coefficients: $\phi$ the angle of maximum gradient, $A$ the gradient on that direction, $B$ the gradient in direction $\phi + \pi/2$, and $c_0$ the nugget variance. The table introduces a feature of cross correlation that does not occur with autocorrelation: The semi-variances can be negative. Figures 17, 18, and 19 show three examples from the possible 15 cross semi-variograms, the three that the authors later used for co-kriging. In two of them the semi-variances are negative.

Table 3 shows that the anisotropy ratio $A/B$ was very similar for sand and silt at both depths, and that the directions of anisotropy were much the

**Table 3.** Coefficients of the Geometric Anisotropic Semi-Variogram for Sand and Silt, Equation [41], Found by Fitting the Model Independently to Each Variable

| | Topsoil | | Subsoil | |
|---|---|---|---|---|
| | Sand | Silt | Sand | Silt |
| Nugget variances, $c_0$ | | | | |
| Topsoil sand | 0 | 0 | 0 | 0 |
| silt | | 0 | 0 | 0 |
| Subsoil sand | | | 0.0675 | −0.0665 |
| silt | | | | 0.0968 |
| Maximum gradients, $A$ | | | | |
| Topsoil sand | 0.0128 | −0.0145 | 0.0198 | −0.0239 |
| silt | | 0.0172 | −0.0230 | 0.0285 |
| Subsoil sand | | | 0.0341 | −0.0418 |
| silt | | | | 0.0547 |
| Minimum gradients, $B$ | | | | |
| Topsoil sand | 0.0080 | −0.0087 | 0.0118 | −0.0138 |
| silt | | 0.0107 | −0.0132 | 0.0159 |
| Subsoil sand | | | 0.0230 | −0.0272 |
| silt | | | | 0.0342 |
| Anisotropy ratios, $A/B$ | | | | |
| Topsoil sand | 1.590 | 1.671 | 1.680 | 1.731 |
| silt | | 1.709 | 1.739 | 1.788 |
| Subsoil sand | | | 1.483 | 1.538 |
| silt | | | | 1.570 |
| | Weighted average 1.590 | | | |
| Directions of maximum variation, | | | | |
| Topsoil sand | 1.148 | 1.147 | 1.166 | 1.188 |
| silt | | 1.151 | 1.176 | 1.208 |
| Subsoil sand | | | 1.165 | 1.163 |
| silt | | | | 1.192 |
| | Weighted average 1.172 | | | |

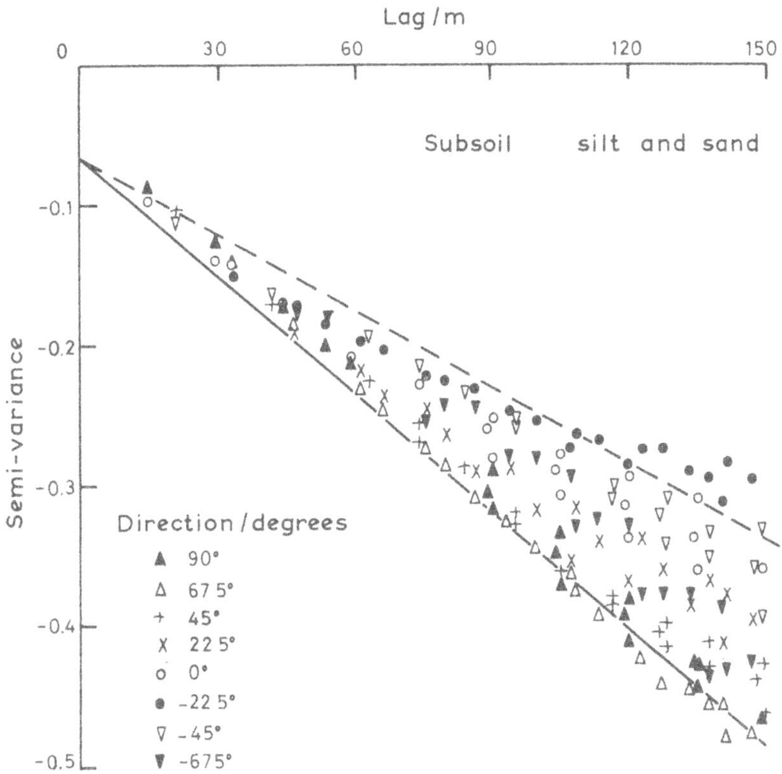


**Figure 17.** Semi-variogram of subsoil silt and sand at Woburn. Here and in Figures 18 and 19 the symbols for the semi-variances show the different directions and the oblique lines the enclosing envelope of the fitted model (see McBratney and Webster, 1983).

same. It seemed that the values obtained were estimates of common values for the two parameters. Weighted averages of them were therefore computed, giving $A/B = 1.59$ and $\phi = 1.17$ rad. With these held constant the model of equation [44] was fitted afresh to give the coefficients listed in Table 4. The authors also presented model coefficients for clay auto semi-variograms. The anisotropy ratios were fairly similar, but the angles of orientation were different, especially that for the subsoil.

## VIII. Changing Drift

Most of the semi-variograms that we have seen so far have had nugget variances, and all have been either linear or convex upward. Occasionally, however, we may encounter semi-variograms of soil properties that

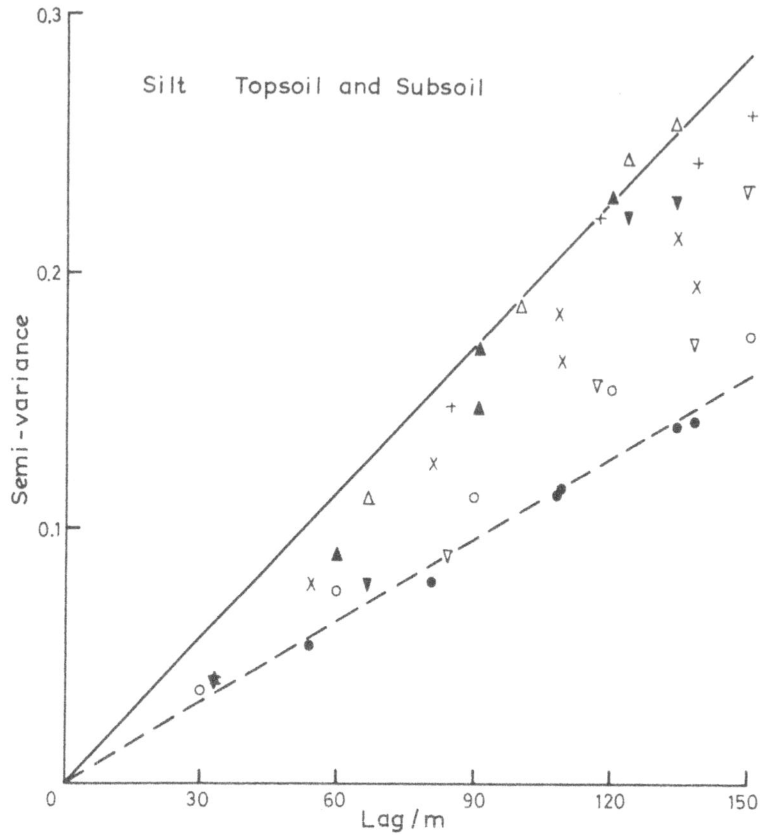


**Figure 18.** Semi-variogram of silt in the topsoil and that in the subsoil at Woburn.

approach the origin and are concave upward, approximately parabolic there. This shape should immediately signal a departure from the assumptions of the intrinsic hypothesis. It is a sign of smooth continuity in the measurements of the soil, so that there are local trends.

Figure 20 shows an example in which the electrical resistivity of the soil was measured at 1-m intervals (Webster and Burgess, 1980), and Figure 21 presents the semi-variograms for the four principal directions of the grid. There are a few places where the resistivity appears fairly constant apart from point-to-point fluctuation; e.g., around 52 m and from 94 m onwards in the upper graph. Elsewhere, however, there are local trends extending over 4 m to about 10 m. The expected value of resistivity is not constant, even within small neighborhoods, but changes with position. It is a function of position, thus

$$\mathrm{E}[z(\mathbf{x})] = m(\mathbf{x}). \qquad [57]$$

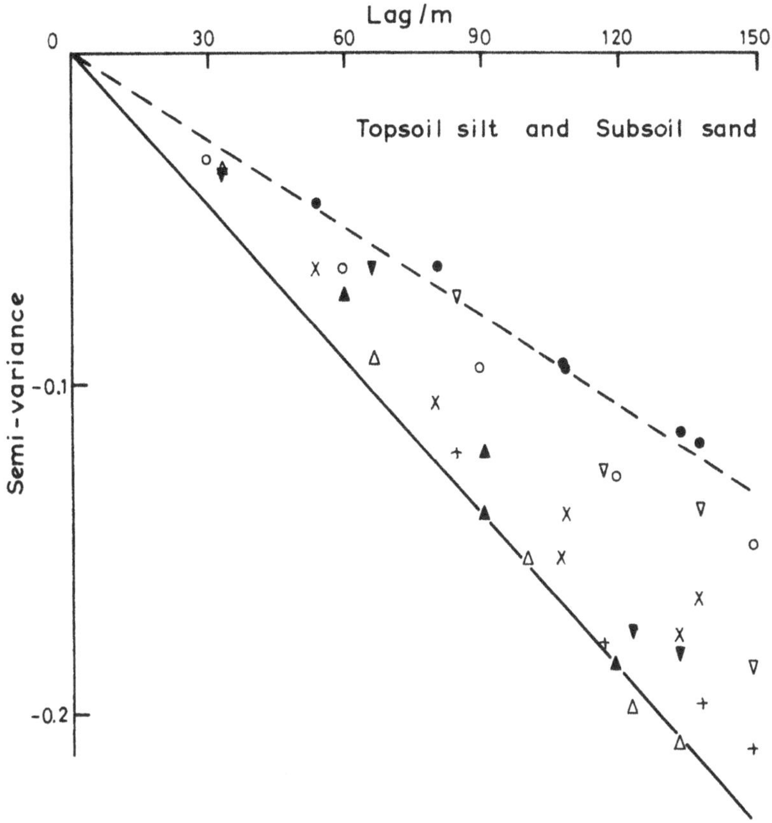


**Figure 19.** Semi-variogram of topsoil silt and subsoil sand at Woburn.

Our model of the soil for the stationary case, equation [7], must be changed by replacing $\mu_v$, the mean, by the more general $m(\mathbf{x})$, giving

$$z(\mathbf{x}) = m(\mathbf{x}) + \varepsilon(\mathbf{x}). \qquad [58]$$

The quantity $m(\mathbf{x})$ representing the trend is known in regionalized variable theory as the *drift*. The term is well entrenched, and despite possible confusion with material transported by ice we shall adhere to it.

### A. Structural Analysis

Where the drift changes, rearranging equation [58] gives $\varepsilon(\mathbf{x})$ as the deviation from the expectation

$$\varepsilon(\mathbf{x}) = z(\mathbf{x}) - m(\mathbf{x}). \qquad [59]$$

In these circumstances equations [6] and [8] are not equivalent. The raw semi-variance no longer estimates the expected squared difference between

**Table 4.** Coefficients of the Geometric Anisotropic Semi-Variogram for Sand and Silt, Equation [41], with Common Anisotropy Ratio, 1.59, and Direction, 1.172 rad

| | Topsoil | | Subsoil | |
|---|---|---|---|---|
| | Sand | Silt | Sand | Silt |
| Nugget variances, $c_0$ | | | | |
| Topsoil sand | 0 | 0 | 0 | 0 |
| silt | | 0 | 0 | 0 |
| Subsoil sand | | | 0.0715 | −0.0668 |
| silt | | | | 0.0998 |
| Maximum gradients, $A$ | | | | |
| Topsoil sand | 0.0128 | −0.0143 | 0.0195 | −0.0233 |
| silt | | 0.0169 | −0.0224 | 0.0275 |
| Minimum gradients, $B$ | | | | |
| Topsoil sand | 0.0080 | −0.0090 | 0.0122 | −0.0146 |
| silt | | 0.0106 | −0.0141 | 0.0173 |
| Subsoil sand | | | 0.0215 | −0.0263 |
| silt | | | | 0.0338 |

the residuals at two places, and we need an analysis that will reveal both the drift and the deviations from it. This is by no means straightforward. Nevertheless, it can be important since it forms the basis of universal kriging.

The aim of the analysis, then, is to estimate the semi-variogram of the residuals, equation [8]. The problem arises because, as equation [59] shows, these residuals depend not only on the measured values, $z(\mathbf{x}_i)$, but also on the drift values, $m(\mathbf{x}_i)$, which we do not know. The problem is compounded because the semi-variogram must be known in order to estimate the drift. Thus, for a given set of data we could compute a semi-variogram if we knew the drifts, which we could calculate if we knew the semi-variogram. This apparent impasse is resolved by a structural analysis that combines trial and error with good judgment. Its solution owes much to Olea (1975) for an English text that leads step by step from the theory to its practical use. Olea (1977) has also provided a computer program for the analysis.

We proceed as follows. We first assume an expression for the drift:

$$m(\mathbf{x}) = \sum_{j=0}^{k} a_j f_j(\mathbf{x}) \qquad [60]$$

where $a_j, j = 0, 1, 2 \ldots k$ are unknown coefficients, and $f_j(\mathbf{x})$ are known

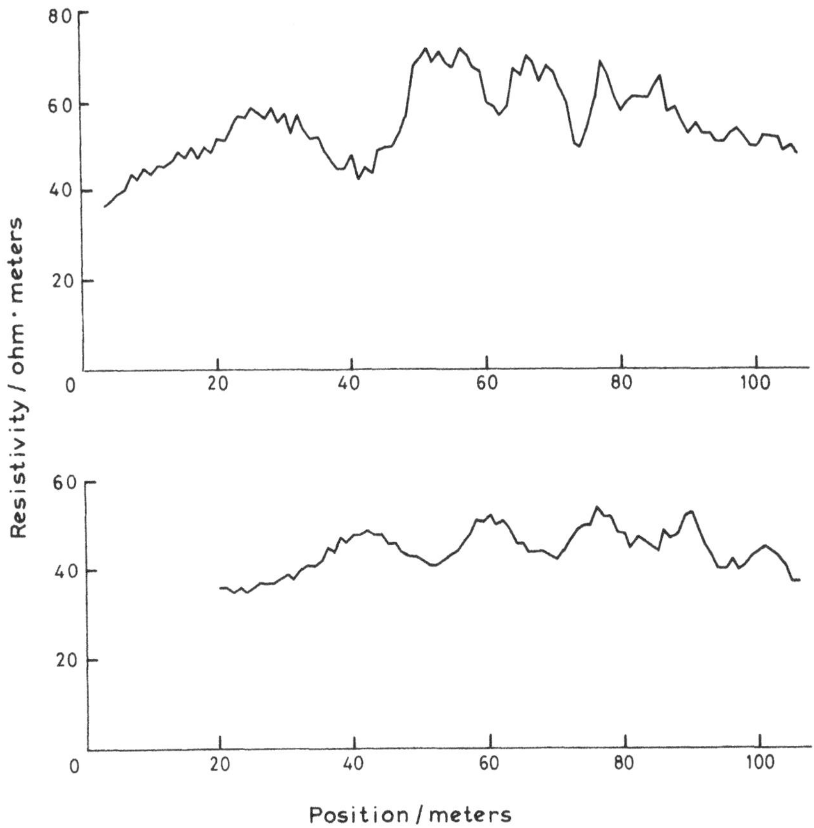


**Figure 20.** Electrical resistivity along two transects at Bekesbourne, England (see Webster and Burgess, 1980).

functions of **x**. We also assume a theoretical semi-variogram and use it to evaluate the coefficients $a_j$ in equation [60]. This gives us value for $m(\mathbf{x})$. Using these estimates of the drift we obtain residuals between them and the data and compute the semi-variogram of the residuals. This should match the theoretical semi-variogram.

Deriving the coefficients in equation [60] is lengthy and need not concern us here: Olea (1975) gives the full details. In practice we can usually keep the task fairly simple by sampling initially on a regular grid, by analyzing one direction at a time, and by restricting the neighborhood over which we consider the semi-variogram as in the stationary case. Given these restraints, the expression for the drift need never be more complex than a second-order polynominal, and the semi-variogram can be well approximated by a straight line,

$$\gamma(h) = wh, \qquad [61]$$

usually without a nugget variance.

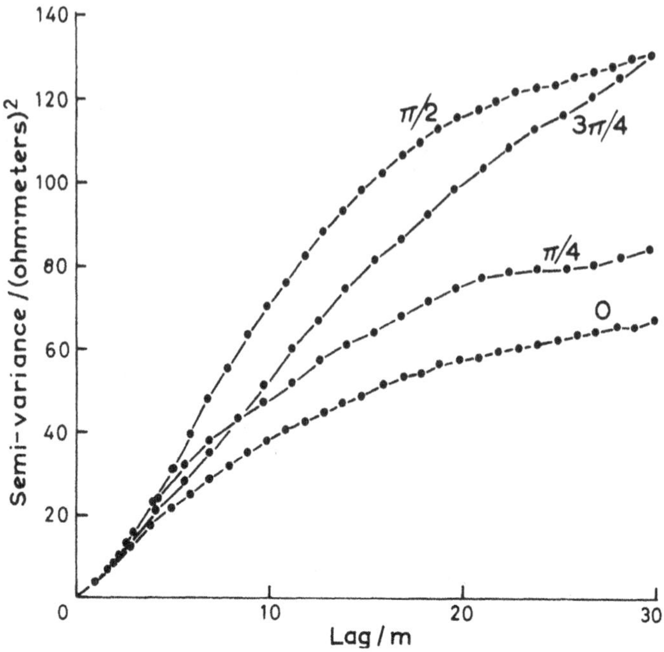


**Figure 21.** Semi-variograms of electrical resistivity in the soil at Bekesbourne, England, for four directions.

Equation [60] for the drift then takes one of the following forms:

$$m(x) = a_0$$
$$m(x) = a_0 + a_1 x$$
or
$$m(x) = a_0 + a_1 x + a_2 x^2, \qquad [62]$$

where $x$ relates to an origin within the neighborhood. These expressions can be combined with a linear semi-variogram that can be changed by lengthening or shortening the neighborhood over which it applies.

The constant $a_0$ cannot be calculated from the semi-variogram, but in any case is not needed since the semi-variogram of the estimated residuals does not depend on it. Olea shows that for regular sampling the other coefficients are given by

$$a_1 = \frac{z(x_n) - z(x_1)}{c(n - 1)}, \qquad [63]$$

for linear drift, and for quadratic drift:

$$a_1 = \frac{z(x_n) - z(x_1)}{c(n - 1)} - (n - 1)\, a_2 c \qquad [64]$$

$$a_2 = -\frac{3[2\bar{z} - z(x_n) - z(x_1)]}{(n - 2)(u - 1)\, c^2}. \qquad [65]$$

Here $x_1$ and $x_n$ are the extreme sampling points in the neighborhood, $c$ is the sampling interval, and $\bar{z}$ is the mean of the sample in the neighborhood; i.e.,

$$\bar{z} = \frac{1}{n} \sum_{i=1}^{n} z(x_i). \qquad [66]$$

When residuals are calculated from either linear or quadratic drift their semi-variances are biased estimates of the semi-variances of the true residuals (Olea, 1975). The bias can be calculated by assuming that the semi-variogram is linear and that its slope is well estimated by the slope of the semi-variogram of the estimated residuals at the origin—effectively the slope over the first lag interval.

For linear drift the slope is

$$w = \frac{n-1}{n-2} \gamma_E(1), \qquad [67]$$

where $\gamma_E(1)$ is the estimated semi-variance at lag 1. The bias at lag $h$ is then given by

$$\frac{wh^2}{c\,(n-1)}, \qquad [68]$$

and the semi-variances of the true residuals are

$$\gamma(h) = \gamma_E(h) + \frac{wh^2}{c\,(n-1)}. \qquad [69]$$

For quadratic drift the slope of the true semi-variogram is estimated from $\gamma_E(1)$ by

$$w = \frac{n-1}{n-3} \gamma_E(1), \qquad [70]$$

and the full semi-variances are then

$$\gamma(h) = \gamma_E(h) + \frac{wh^2[c^2(2n^2 - 2n - 1) - 2cnh + h^2]}{(n-2)(n-1)\,nc^3}. \qquad [71]$$

A choice is first made of combination of drift power and neighborhood size, and the analysis is performed. If the experimental semi-variogram matches the theoretical one well then the combination is judged an appropriate model to describe the variation of the soil property in the region. If, however, the match is poor then either the expression for the drift or the size of neighborhood or both are changed and the analysis is repeated. It is not always easy to decide whether a particular match is good, and even if the match seems good there might be better ones. So the analysis is repeated with successively larger neighborhoods, first for linear drift and then for quadratic drifts, and the results compared. The best fit is chosen by inspecting graphs of the results, and there may be more than one equally good combination.

Webster and Burgess (1980) performed a full structural analysis on the resistivity data whose raw semi-variogram is shown in Figure 21. Figures 22 to 25 show examples for neighborhoods of just five and nine terms, equal to 4 and 8 lag intervals. Note, however, that an interval across the diagonals is $\sqrt{2}$ times that along the rows or columns of the grid. In the figures the circles represent the assumed linear semi-variogram and the stars the actual semi-variograms of the residuals.

As might be expected the agreement between assumed and actual semi-variograms is good for both linear and quadratic drifts when the

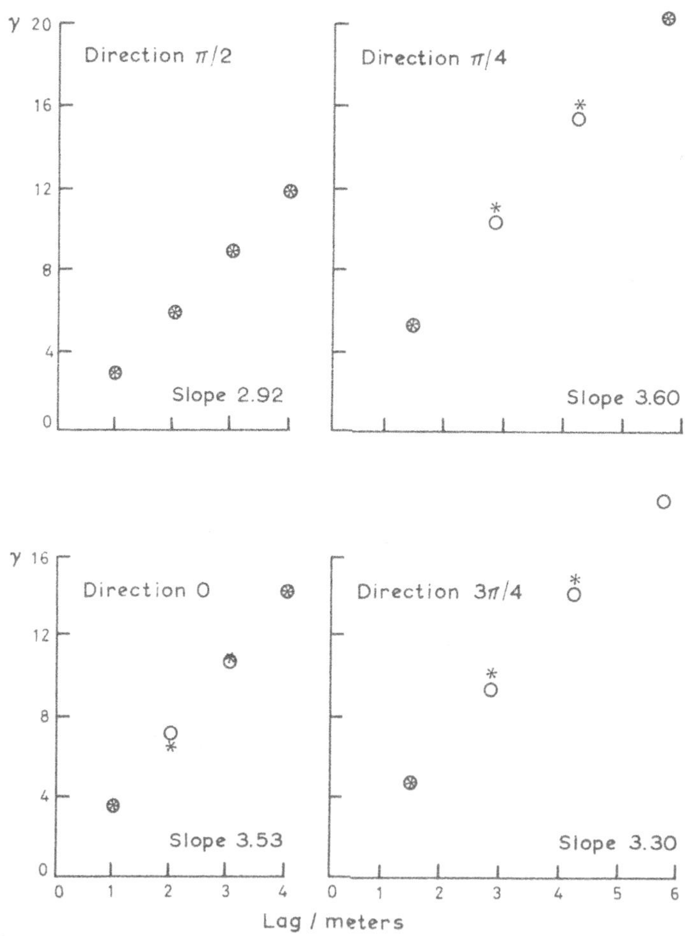


**Figure 22.** Semi-variograms of residuals from linear drift over five terms (four lag intervals) at Bekesbourne. Here and in Figures 22 to 24 the ordinate is in (ohm. meters)$^2$ and the lag in meters.

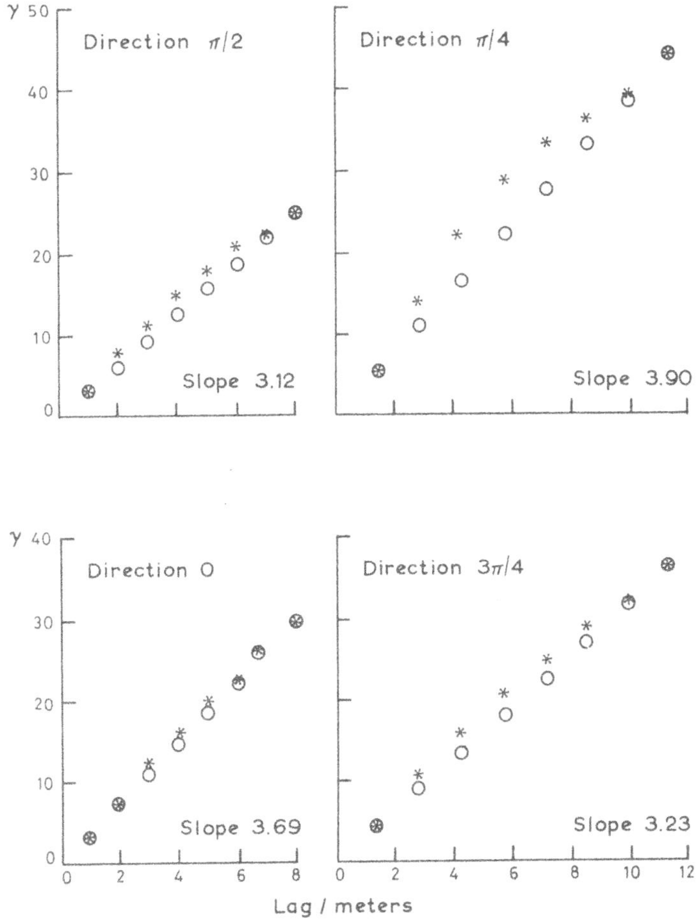


**Figure 23.** Semi-variograms of residuals from linear drift over nine terms.

neighborhood is restricted to five terms. When enlarged to nine terms the agreement is still good for the quadratic drift, but somewhat less so for the linear drift, especially in the north-south direction. Webster and Burgess found that on increasing the drift from these the agreement became steadily worse. Further, because they wished to use the semi-variogram for kriging (see later) across large gaps in their grid of data they chose the linear model in preference to the quadratic because the latter was inadequately constrained there.

On fitting equation [43] for geometric anisotropy to the semi-variograms from the linear drift over nine terms the following is obtained:

$$\gamma(h, \theta) = [3.913^2 \cos^2(\theta - 0.416) + 3.029^2 \sin^2(\theta - 0.416)]^{1/2} |\mathbf{h}|, \quad [72]$$

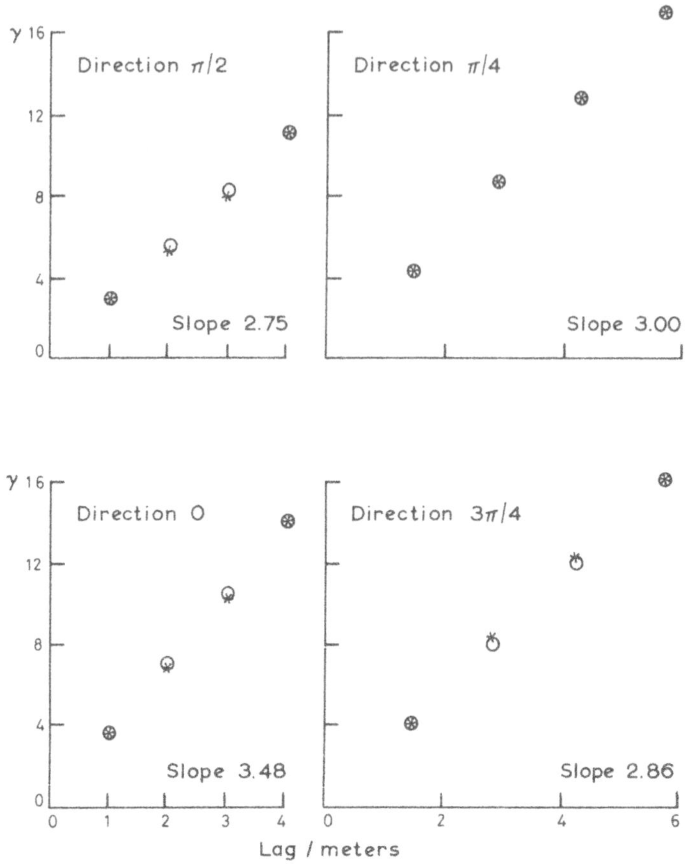


**Figure 24.** Semi-variograms of residuals from quadratic drift over five terms.

**Table 5.** Slopes in (ohm m)$^2$ m$^{-1}$ of Semi-Variograms Assuming Constant Drift and of Residuals from Linear and Quadratic Drift

| Direction | Constant drift average slope | | Linear drift slope | | Quadratic drift slope | |
|---|---|---|---|---|---|---|
| | 5 terms | 9 terms | 5 terms | 9 terms | 5 terms | 9 terms |
| NW to SE | 5.82 | 6.88 | 2.92 | 3.12 | 2.75 | 2.73 |
| SW to NE | 4.42 | 4.03 | 3.53 | 3.69 | 3.48 | 3.49 |
| N to S | 5.59 | 4.60 | 3.60 | 3.90 | 3.00 | 3.45 |
| W to E | 4.91 | 5.26 | 3.30 | 3.23 | 2.86 | 3.05 |

where $h = |\mathbf{h}|$ is in meters and $\theta$ is in radians. Table 5 compares the calculated slopes, $w$, and values given by the fitted model, and shows good agreement between the two.

## IX. Extension to the Power Spectrum

The semi-variogram and the co-variogram both express the spatial dependence of a property in the spatial domain. In time series analysis they express variation in the domain of time. Many temporal phenomena are periodic, and the semi-variogram will itself then show periodicity. In these cases it can be helpful to transform the semi-variogram, or better the co-variogram or correlogram into the frequency domain by Fourier analysis.

The transformed function is the *power spectrum*, or simply the *spectrum*. It expresses the distribution of variance with frequency. By definition the spectrum is given by

$$\Gamma(f) = \lim_{n\to\infty} \mathrm{E}[G(f)] = \int_{-\infty}^{\infty} C(h)e^{-j2\pi fh}dh, \qquad [73]$$

where $f$ denotes the frequency, $G$ is the spectral density for a finite series of length $n$, and $j$ is $\sqrt{-1}$. If $C(h)$ is replaced by the autocorrelation then the transform gives the spectral density, which is

$$\mathrm{P} = \frac{\Gamma(f)}{\sigma^2} = \int_{-\infty}^{\infty} \rho(h)e^{-j2\pi fh}dh, \qquad [74]$$

and this, like the autocorrelation, no longer depends on the scale of measurement.

Estimating the spectrum is a large subject in its own right and still a matter of some controversy. Readers who wish to use the technique should consult one or more of the standard works by, for example, Chatfield (1980), Jenkins and Watts (1968), and Kendall and Stuart (1976). Basically, the required transform of the co-variogram is

$$\hat{\Gamma}(f) = \frac{1}{A}\left[\hat{C}(0) + 2\sum_{k=1}^{m} \hat{C}(i)\cos \pi kf\right], \qquad [75]$$

with the generally agreed value for $A$ of $2\pi$. The value of $m$ is the maximum lag over which the summation is performed. Estimates so obtained, however, are subject to substantial error, and therefore they are usually smoothed. There are several methods for this. They involve viewing the

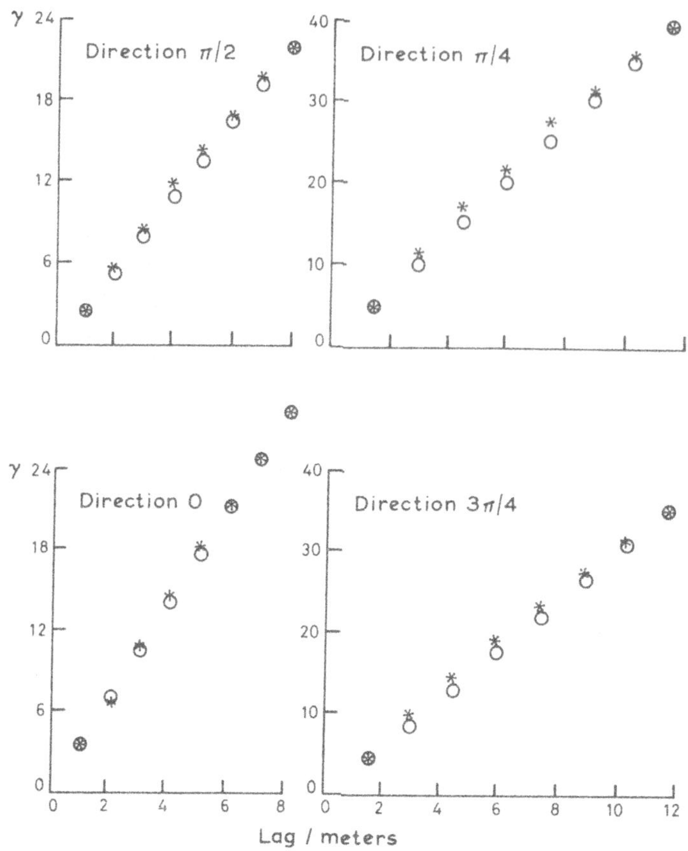


**Figure 25.** Semi-variogram of residuals from quadratic drift over nine terms.

spectrum through a *window*, the width of which can be chosen as desired: The wider the window the greater is the smoothing, though the larger becomes the risk of bias in the estimates. The ideal width is the one that achieves the best compromise between precision and accuracy in the estimates. The shape of the window can also be varied.

An example to illustrate this extension of spatial analysis is taken from Webster (1977). A transect had been sampled at 4-m intervals across gilgai terrain on the Bland Plain in eastern Australia, and at each sampling point several soil properties had been recorded. Among these was the electrical conductivity at 30- to 40-cm depth, measured in a 1:5 soil:water extract. There was a total of 355 observations over 1420 m. The measured values were strongly skewed, and they were therefore transformed to logarithms to stabilize the variances, and on this scale the series seemed fully second-order stationary.

The correlogram was computed. It is shown as its complement, the semi-variogram, in Figure 27. It was then transformed to its spectrum using the Bartlett window as smoothing function (Bartlett, 1966). This function is defined for the frequency domain as

$$W(f) = m\left(\frac{\sin fm}{fm}\right)^2, \qquad [76]$$

where $m$ is the maximum lag of the transformation. In computing, however, it is usually applied to the correlogram in the spatial or time domain as a *lag window*, $w(f)$, which is simply

$$\begin{aligned} w(f) &= 1 - k/m \qquad && \text{for } k \leq m \\ w(f) &= 0 && \text{for } k > m. \end{aligned} \qquad [77]$$

This then gives the computing formula for spectral density as

$$R(f) = \frac{1}{2\pi}\left[1 + 2\sum_{k=1}^{m} \hat{\rho}(k)\, w(k) \cos\frac{\pi f k}{L}\right], \qquad [78]$$

where $f$ is the frequency ranging from 0 to ½ cycle and $L$ is the number of steps by which $f$ is incremented. A half cycle corresponds to two sampling intervals and is the limit, often known as the *Nyquist* frequency. If the spectrum is needed for shorter wavelengths, then sampling must be more intensive. In this study the frequency therefore ranged from 0 to 1/8 cycles per meter (wavelength 8 m) and $L$ was set to 100.

The spectrum was computed with $m$ set to four different sampling intervals, 10, 25, 40, and 60. Figure 26 shows the results. With a lag window of 10 the spectrum declines smoothly over the frequency range. As $m$ is increased, however, more detail becomes apparent, and in particular a strong peak emerges at 0.12 cycles. This frequency, 0.12 cycles per 4 m, corresponds to a wavelength of 33.3 m, or a little more than eight sampling intervals. It signifies periodic fluctuation in the conductivity with this wavelength, a result that matched the morphological evidence.

In the original study no attempt was made to model the correlogram, and the semi-variogram was not computed. In the light of the foregoing sections it is worth seeing how the modern geostatistical approach treats such periodicity. The conductivities in their logarithmic form are stationary, and so the sample semi-variogram is a mirror image of the co-variogram. It too is periodic, therefore, and some kind of angular function is needed as a model of it. Journel and Huijbregts (1978) suggest some appropriate models and their limiting values to ensure that the semi-variograms are positive definite. In this instance an angular function alone is not sufficient because the semi-variogram clearly increases from a small value at lag 1 to much larger values, reminiscent of a sill about which it thereafter fluctuates.

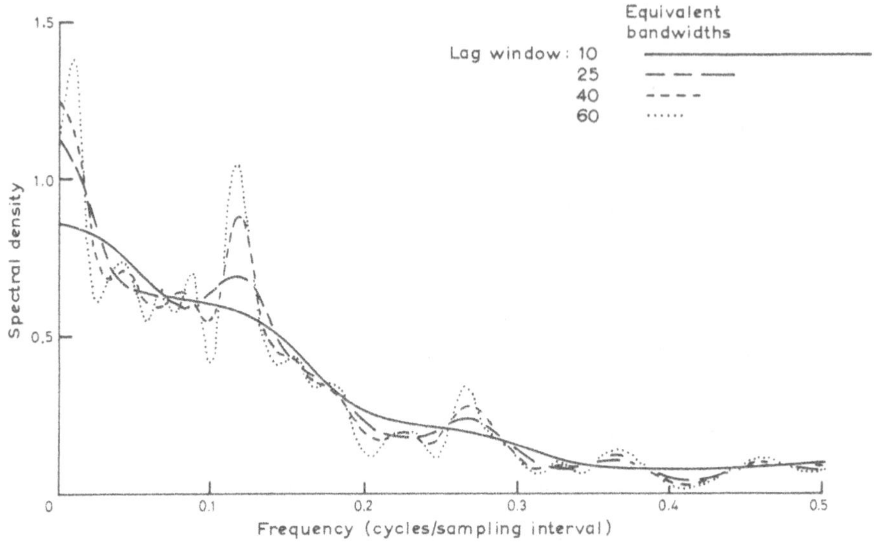


**Figure 26.** The spectrum of electrical conductivity of the soil at 30 to 40 cm on the Bland Plain computed with Bartlett lag windows of four widths (see Webster, 1977).

This additional component seemed best modeled by an exponential term. The best fitting model finally chosen was

$$\gamma(h) = c + b \exp(-h/a) + u \sin\left(\frac{2\pi h}{l}\right) + v \cos\left(\frac{2\pi h}{l}\right) \quad \text{for } h > 0$$
$$\gamma(0) = 0, \qquad [79]$$

with its six parameters estimated by

$$\hat{a} = 0.723 \qquad \hat{u} = 0.00937$$
$$b = 0.220 \qquad \hat{v} = 0.00833$$
$$\hat{c} = 0.188 \qquad l = 8.89.$$

Figure 27 shows the sample semi-variances and the fitted function. Note especially the estimated wavelength, $l$, of 8.89 sampling intervals, which is sensibly the same as that found by spectral analysis.

In geostatistical parlance any semi-variogram whose increase is not monotonic is said to exhibit a "hole effect." This signifies some kind of repetition in the regional distribution of the variable. It was evident in Webster and Cuanalo's (1975) original correlograms of the soil across the Jurassic limestones, sandstones, and clays of north Oxfordshire when extended to lags of 600 m. McBratney and Webster (1981a) found similar hole effects in some semi-variograms, again to 600 m, of soil on the drift-covered Buchan Plateau in northeast Scotland. Where the periodicity is as strong as it is at Caragabal on the Bland Plain, however, a more

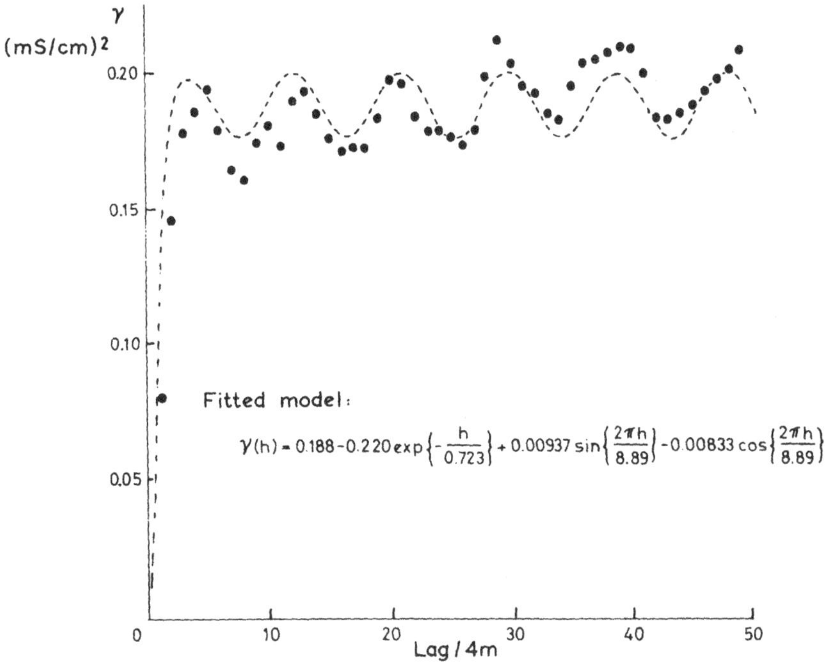


**Figure 27.** The semi-variogram of electrical conductivity of the soil at 30 to 40 cm on the Bland Plain with a periodic model fitted.

connotative term is desirable, and I suggest that such semi-variograms be described as periodic.

## X. Optimal Estimation—Kriging

One of the prime reasons for obtaining a semi-variogram is to use it for estimation, and no review of modern spatial analysis would be complete without some account of how this is done. In soil survey we can recognize two main kinds of estimate. In the one we may wish to estimate the average value of a soil property within some defined region. The other is interpolation: We may wish to predict the values of a property at places that we have not visited. If such places are few we could go there and measure them, of course. In principle, however, they are infinite, and we may wish to predict the values at any or all of them and to make an isarithmic ("contour") map of the property.

If the region of interest is sufficiently uniform then the mean of an unbiased sample estimates both the population mean and the value at any one place. Usually it is not, and in these circumstances surveyors classify the soil so that they can use class means as predictors for individual sites.

Their model is that the value of a property $z$ at a place $i$ in class $j$ is the sum of three terms:

$$z_{ij} = \mu + \alpha_j + \varepsilon_{ij}, \qquad [80]$$

where $\mu$ is the general mean of the property in the whole region, $\alpha_j$ is the difference between the general mean and the mean of class $j$, and $\varepsilon_{ij}$ is a random component with zero mean and variance $\sigma_w^2$, usually assumed to be normally distributed. This is the model underlying the sampling studies by Thornburn and co-workers in Illinois (Thornburn and Larsen, 1959; Morse and Thornburn, 1961) and the research on spatial prediction by Beckett's group at Oxford (e.g., Webster and Beckett, 1968; Beckett and Burrough, 1971).

Following classical procedure class means are estimated by sampling each class and computing

$$\bar{z}_j = \hat{\mu}_j = \frac{1}{n_j}\sum_{i=1}^{n_j} z_{ij}. \qquad [81]$$

Associated with $\hat{\mu}_j$ is its standard error or the square, the estimation variance. The latter is defined as

$$\sigma_{E_j}^2 = E[(\mu_j - \hat{\mu}_j)^2] = \sigma_{w_j}^2/n, \qquad [82]$$

and estimated by

$$\hat{\sigma}_{E_j}^2 = s_{E_j}^2 = \frac{1}{n_j(n_j - 1)}\sum_{i=1}^{n_j} (z_{ij} - \bar{z}_j)^2. \qquad [83]$$

If the soil has been classified at the same categoric level throughout the region, then the variances within all the classes should be the same, $\sigma_{w_j}^2 = \sigma_w^2$.

The pooled within-class variance should therefore be well estimated from a large sample.

The estimation variance for predicting a point, assuming the class is known, is the within-class variance plus the estimation variance for the mean of that class, thus:

$$\sigma_{Ep}^2 = \sigma_w^2 + \sigma_w^2/n_j. \qquad [84]$$

We note that in classical procedure the errors associated with prediction or interpolation are determined very largely by the size of the within-class variance: $\sigma_{Ep}^2$ cannot be less than $\sigma_w^2$, no matter how intensive the sampling. The quality of the classification is paramount, and this is the statistical reason for devoting so much effort to creating a good classification in traditional practice.

The dependence on classification for improving prediction and the inability to take into account either the intensity of sampling or its configuration are serious weaknesses of traditional practice. Regionalized variable theory provides an alternative, and we devote the remainder of this review to it. It is a large subject, and we can examine only the broad principles and some of its simpler forms.

### A. Kriging Defined

The method of estimation embodied in regionalized variable theory is known in earth sciences as *kriging*, after D.G. Krige, who devised it empirically for use in the South African goldfields (see Krige, 1966). It is essentially a means of *weighted local averaging* in which the weights are chosen so as to give unbiased estimates while at the same time minimizing the estimation variance. Kriging is in this sense optimal.

Consider a typical situation in which we have measured a property at a number of places, $\mathbf{x}_i$, within a region to give values $z(\mathbf{x}_i), i = 1, 2 \ldots n$. From these we wish to estimate the value of the property at some place $B$. The place might be a "point," that is, an area of the same size and shape as those on which the measurements were made. Alternatively it might be a larger area or block. The procedures for estimation in the two instances are known as *punctual kriging* and *block kriging*, respectively. Punctual kriging can be regarded as a special case of block kriging, and we shall treat it so here. In either event our estimate at $B$ is the linear sum

$$\hat{z}(B) = \lambda_i z(\mathbf{x}_1) + \lambda_2 z(\mathbf{x}_2) + \ldots + \lambda_n z(\mathbf{x}_n), \qquad [85]$$

where the $\lambda_i$ are the weights associated with the sampling points.

We want our estimate to be unbiased; i.e.,

$$E[(z(B) - \hat{z}(B)] = 0, \qquad [86]$$

and this is assured if the weights sum to 1:

$$\sum_{i=1}^{n} \lambda_i = 1. \qquad [87]$$

The estimation variance at $B$ is the expected square difference between our estimate and the true value, and is

$$\sigma_E^2(B) = E\{[z(B) - \hat{z}(B)]^2\} = 2 \sum_{i=1}^{n} \lambda_i \bar{\gamma}(\mathbf{x}_i, B)$$
$$- \sum_{i=1}^{n} \sum_{j=1}^{n} \lambda_i \lambda_j \gamma(\mathbf{x}_i, \mathbf{x}_j) - \bar{\gamma}(B, B). \qquad [88]$$

Here $\gamma(\mathbf{x}_i, \mathbf{x}_j)$ is the semi-variance of the property between $\mathbf{x}_i$ and $\mathbf{x}_j$, taking account of both the distance, $\mathbf{x}_i - \mathbf{x}_j$, separating them and the angle, $\bar{\gamma}(\mathbf{x}_i, B)$ is the average semi-variance between $\mathbf{x}_i$ and all points within the block, and $\bar{\gamma}(B, B)$ is the average semi-variance within the block; i.e., the within-block variance defined as in equation [48] earlier. In punctual kriging this last term is zero—there is no variance at a point—and the quantity $\bar{\gamma}(\mathbf{x}_i, B)$ is just the semi-variance between $\mathbf{x}_i$ and the point to be estimated.

The estimation variance is minimized, subject to the constraint that $\lambda_i$ = 1, by finding the partial derivatives with respect to each $\lambda_i$, and introduces a Lagrange multiplier, $\psi$. The minimum is obtained when

$$\sum_{i=1}^{n} \lambda_i\, \gamma(\mathrm{x}_i, \mathrm{x}_j) + \psi = \bar{\gamma}(\mathrm{x}_j, B) \qquad \text{for } j = 1, 2, \ldots n. \qquad [89]$$

There are thus $n$ equations in $n$ unknowns plus a set of unbias conditions from which to determine $\psi$. These are best represented in matrix form by

$$\mathbf{A} \begin{bmatrix} \lambda \\ \psi \end{bmatrix} = \mathbf{b}, \qquad [90]$$

where

$$\mathbf{A} = \begin{bmatrix} \gamma(\mathrm{x}_1, \mathrm{x}_1) & \gamma(\mathrm{x}_1, \mathrm{x}_2) & \ldots & \gamma(\mathrm{x}_1, \mathrm{x}_2) & 1 \\ \gamma(\mathrm{x}_2, \mathrm{x}_1) & \gamma(\mathrm{x}_2, \mathrm{x}_2) & \ldots & \gamma(\mathrm{x}_2, \mathrm{x}_1) & 1 \\ \vdots & \vdots & & \vdots & \vdots \\ \gamma(\mathrm{x}_n, \mathrm{x}_1) & \gamma(\mathrm{x}_n, \mathrm{x}_2) & \ldots & \gamma(\mathrm{x}_n, \mathrm{x}_n) & 1 \\ 1 & 1 & \ldots & 1 & 0 \end{bmatrix},$$

$$\begin{bmatrix} \lambda \\ \psi \end{bmatrix} = \begin{bmatrix} \lambda_1 \\ \lambda_2 \\ \vdots \\ \lambda_n \\ \psi \end{bmatrix} \text{ and } \mathbf{b} = \begin{bmatrix} \gamma(\mathrm{x}_1, B) \\ \gamma(\mathrm{x}_2, B) \\ \vdots \\ \lambda(\mathrm{x}_n, B) \\ 1 \end{bmatrix}. \qquad [91]$$

Having solved equation [90] to find the weights, the estimation variance is obtained from

$$\sigma_E^2 = \mathbf{b}^T \begin{bmatrix} \lambda \\ \psi \end{bmatrix} - \bar{\gamma}(B, B). \qquad [92]$$

Kriging thus provides not only unbiased estimates of minimum variance, but also a measure of the estimation variance. In this respect it is superior to other methods of interpolation.

### B. Example

Applying these equations to the stone content on Plas Gogerddan will reveal several interesting features of kriging. The semi-variances are taken from equation [44].

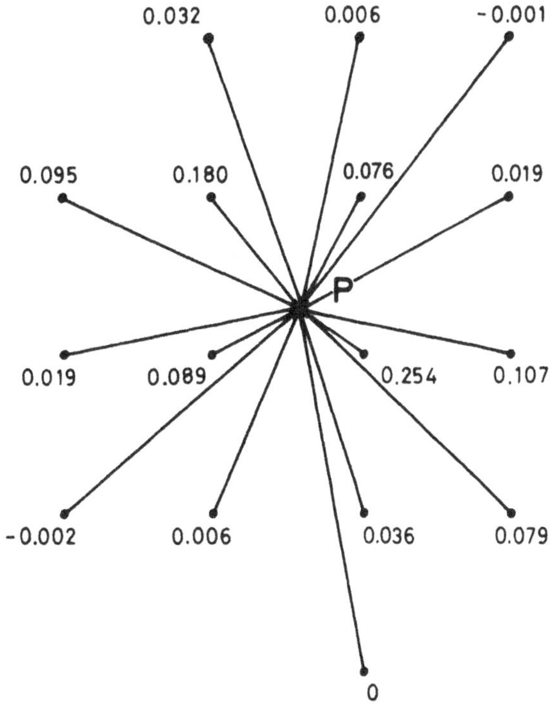


**Figure 28.** Weights for kriging stone content at a point, *P*, at Plas Gogerddan.

Suppose we wish to estimate the stone content at a point *P*. Figure 28 shows *P* lying one-third of a sampling interval from the bottom and right-hand side of a grid cell. The sampling points are at the nodes of the grid, and the nearest 16 are shown. The values at the nodes are the weights. The nearest is much the largest, and the ones in the outer shell are much smaller in absolute value. Notice especially that those in the bottom left and top right corners are so small as to be negligible. The figure also shows the effect of anisotropy. Variation is greatest approximately in the direction of these points and least in the perpendicular direction. As a result points lying in the latter direction carry more weight than those in the former for the same distance.

The particular weights depend of course on the semi-variogram, especially the size of the nugget variance, and on the configuration of sampling points and place to be predicted. Nevertheless, the fact that near data carry much weight and far data none of consequence is quite general and very important. It means that kriging is essentially local, and conforms to our intuitive notion of a sensible estimation procedure. It also means that for kriging the semi-variogram needs to be well estimated only to the lag of the furthest point to carry effective weight. In many instances the semi-variogram will be linear to this lag, and so fitting a model is easy (see Figure 13). Of course, there must be ample data within the range of the semi-variogram if it is transitive, since any beyond it are spatially independent of the point being estimated. Local estimation, or interpolation, is not worth attempting with spatially independent data.

The local nature of kriging is computationally important, for it means that we rarely need more than 16 observations from which to krige one estimate, and matrix **A** need not be larger than 17 × 17. This does not matter when kriging just one point or block, but to make a map might require thousands of matrix inversions if the data are irregularly scattered. The size of the matrix is then very serious.

Applying the procedure to the whole of the field at Plas Gogerddan gives the results shown in Figures 29 to 33. Figure 29 is an isarithmic map drawn by interpolating a 3 × 3 square grid of points within each original grid cell and then threading contour lines through the grid. Notice in particular the discontinuity around many of the original grid nodes. These are due to the nugget effect and the minimum variance criterion. The latter ensures that the interpolated value at a sampling point is the observed value there. The presence of a nugget variance means that the semi-variogram is effectively composed of two distinct functions, one, $\gamma_1(h, \theta) = w(\theta)\, h$, describing the spatial dependence, and the other the purely random function, $\gamma_0 = c_0$, that represents the discontinuity. Its effect appears even more dramatically in Figure 30, a block diagram of the interpolated surface.

A user of a map will often want to know not about points but about small blocks, of, say, 1 to 10 ha. In these circumstances a block kriged map should be made. Figure 31 shows such a map made from the same data. Each block was 30.5 × 30.5 m, and again there were nine estimates for each original grid cell on a 5.1-m mesh. The result is smoother than the punctually kriged map. Note, however, that the aim is not primarily cosmetic: it is to provide the best estimates for 30.5 × 30.5-m blocks.

As mentioned above, kriging provides estimates of the estimation variances. These themselves can be displayed in the form of a reliability map. Where both sampling and interpolation are on regular grids, as here, the estimation variances have a regular pattern that is repeated. Figure 32 gives the estimation variances for grid cells in the central part of the map, (a) for punctual kriging and (b) for block kriging. A block diagram of the block kriging variance for the whole field, Figure 33, is interesting in emphasizing the way in which the error increases sharply near the edge of the field, beyond which there are no data. The small bump near the bottom is caused by a missing value.

The above is just one example to illustrate the power and reliability of kriging for local estimation and mapping. There are now a number of examples in the literature of soil science (Burgess and Webster, 1980a, 1980b; McBratney *et al.*, 1982; Yost *et al.*, 1982; Vieira *et al.*, 1981; McBratney and Webster, 1983).

Before leaving the topic we should be clear that a kriged map of a soil property can be a misleading representation of reality in one respect. Even a

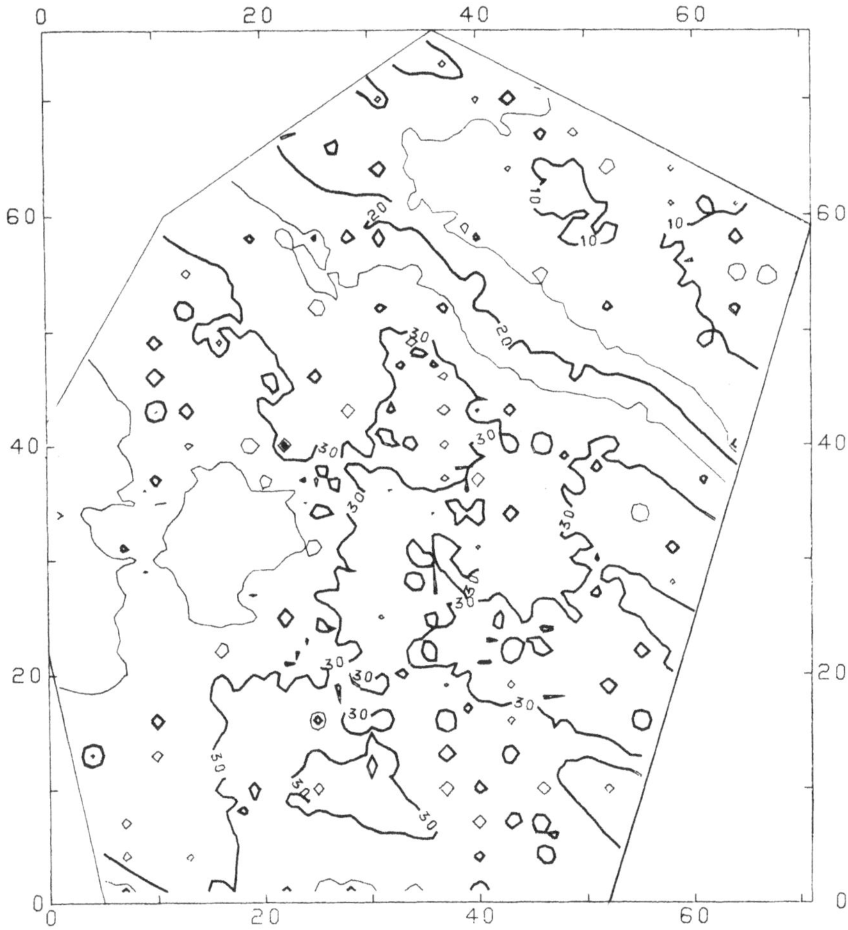


**Figure 29.** Isarithmic map of stone content made by punctual kriging at Plas Gogerddan.

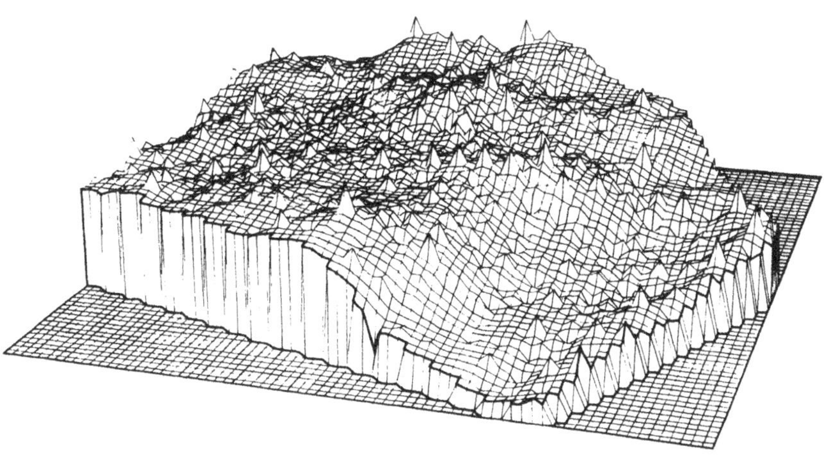

**Figure 30.** Block diagram of stone content estimated by punctual kriging.

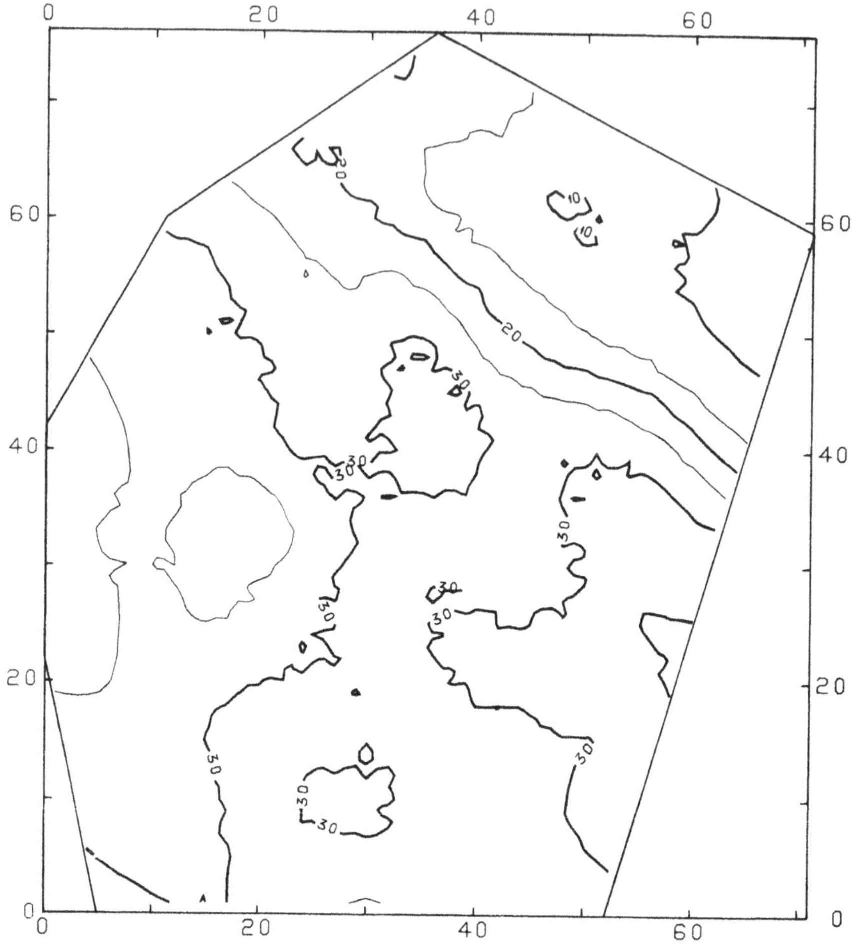


**Figure 31.** Isarithmic map of stone content made by block kriging.

punctually kriged map shows smooth variation between sampling points. Yet if there is a nugget variance, then we know that the soil does not vary smoothly at our working scale. If we wish to represent the soil with all its variation from sample data then we should simulate it in a way that retains all the characteristics of the semi-variogram. Kriging, by averaging, does not do this. Readers who wish to follow the matter further should consult Journel and Huijbregts (1978).

## XI. Designing Sampling Schemes

A second way in which the results of spatial analysis can be used is in designing optimal-sampling schemes for surveys of soil properties. This potential application of regionalized variables theory to soil survey is due

Punctual kriging

| | | | |
|---|---|---|---|
| ● | ○ | ○ | ● |
| 0 | 12.89 | 12.89 | 0 |
| ○ | ○ | ○ | ○ |
| 12.89 | 12.87 | 12.87 | 12.89 |
| ○ | ○ | ○ | ○ |
| 12.89 | 12.87 | 12.87 | 12.89 |
| ● | ○ | ○ | ● |
| 0 | 12.89 | 12.89 | 0 |

Block kriging

| | | | |
|---|---|---|---|
| ● | ○ | ○ | ● |
| 0.929 | 0.935 | 0.935 | 0.929 |
| ○ | ○ | ○ | ○ |
| 0.935 | 0.940 | 0.940 | 0.935 |
| ○ | ○ | ○ | ○ |
| 0.935 | 0.940 | 0.940 | 0.953 |
| ● | ○ | ○ | ● |
| 0.929 | 0.935 | 0.935 | 0.929 |

**Figure 32.** Variances for punctual and block kriging of stone content. Values are in (percent)$^2$. For punctual kriging the block discs are sampling points and open circles the interpolated ones. In the lower figure they represent the centers of the estimated blocks.

largely to Burgess *et al.* (1981) and McBratney *et al.* (1981), and this account is taken largely from their work.

To understand the basis of this application consider equations [88] and [92] for the kriging variance. We can see from these that the errors of estimation depend on the semi-variogram and through it on the configuration of sampling points in relation to the block to be estimated. They do not depend on the observed values themselves. So, if the semi-variogram is known then the kriging variances can be estimated before performing the sampling.

The kriging variance is not constant from place to place as Figure 32 shows. Rather, it tends to increase the further the estimated point or block is from the data, at least where the semi-variogram is a monotonic increasing function, which is usually so for soil. The kriging variance can be diminished on average by sampling more intensively. For any given

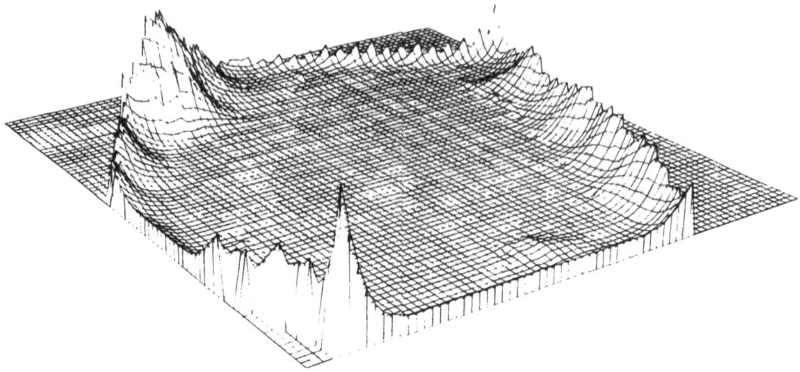

**Figure 33.** Diagram of the block kriging variance.

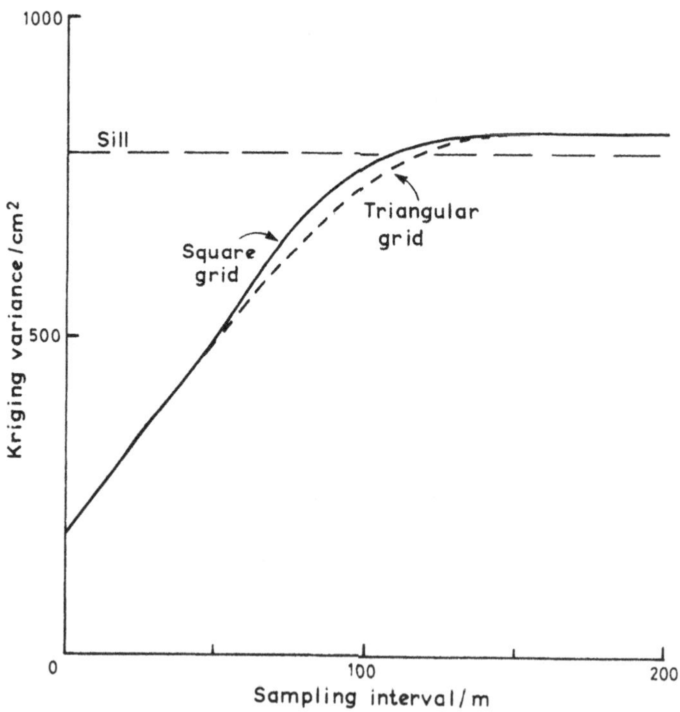


**Figure 34.** Graphs of maximum variance for punctual kriging of thickness of cover loam against sample spacing at Hole Farm, Norfolk, on square and triangular grids.

intensity, however, the maximum kriging variance will still occur where the data are furthest from the estimated point. One criterion of a good sampling scheme is that it minimizes the maximum kriging variance, say, $\sigma^2_{k_{max}}$. If variation is isotropic then it can be achieved by sampling on a regular equilateral triangular grid. For unit sampling density, the side of the triangle is 1.0746 and the maximum distance, $d_{max}$, between a sampling point and a point to be estimated, at the center of a triangle, is 0.6304. Any other sampling scheme, especially random and other irregular schemes, will have some larger values of $d_{max}$ and therefore larger maximum kriging variances. Nevertheless, a square grid with $d_{max} = 1/\sqrt{2} = 0.7071$ for unit sampling density will generally give $\sigma^2_{k_{max}}$ only a little larger than than that for the triangular scheme, and for reasons of convenience in indexing computer management, site location and logistics will usually be preferred.

In planning a survey this information may be used in one of two ways. If the budget for the survey is fixed then so is the sampling intensity. The maximum estimation variance is determined by solving equation [92]. For punctual estimates it is solved just once for the centers of the grid cells. For block estimates, however, $\sigma^2_{k_{max}}$ can occur where the block is centered over a grid node: It depends on the size of the blocks in relation to the grid mesh. The equation must therefore be solved twice to find the maximum.

If, however, funding depends on the precision required, then a maximum tolerable error may be specified for the survey. In such a situation the aim is to determine the sampling intensity that provides just the required precision. The aim is achieved by solving equation [92] for a range of spacings. A graph of $\sigma^2_{k_{max}}$ against sample spacing is plotted and the required spacing read from the graph. The survey can then be performed most economically by sampling at this intensity.

An example from Burgess *et al.* (1981) illustrates the procedure where variation is isotropic. It derives from a study at Hole Farm in eastern England where crop performance had been found to depend on the thickness of cover loam. In one field of 18 ha the thickness had been measured by boring on a 20 m square grid to give some 450 observations. Semi-variances were calculated to 300 m and an isotropic spherical model, equation [25], with the following coefficients fitted:

$$\text{nugget variance } c = 187.0 \text{ cm}^2$$
$$\text{sill-nugget } c = 603.8 \text{ cm}^2$$
$$\text{range } a = 101.2 \text{ m.}$$

Figure 34 shows the graph of maximum kriging variance for punctual estimates against sample spacing for square grids. For comparison the graph for the equilateral triangular configuration is also plotted, and shows that there could be a small gain in efficiency using it. Note that the kriging variance cannot be less than the nugget variance, however dense the sampling, and that it reaches a maximum somewhat greater than the sill value. This maximum is in fact the sill value plus the Lagrange multiplier, $\psi$, of equation [90]. The spacing at which it is reached is $\sqrt{2}$ times the range of the semi-variogram for the square grid. At this spacing the interpolated point at the center of a grid cell becomes independent of the data. The semi-variances in equation [90] equal the sill and all weights are equal to $1/n$. The Lagrange multiplier takes the value $\sigma^2/n$ and represents the additional uncertainty in estimating the value at a place from local data rather than the whole population, and in these circumstances equation [92] is equivalent to the classical equation [84].

Examples of graphs for block kriging from the same semi-variogram are shown in Figure 35 for two sizes of block. As above, two graphs can be drawn, one for blocks whose centers lie in the centers of the grid cells and the other for blocks centered over grid nodes. At most spacings the kriging variances are larger for the former, but there is a small range of spacings where they are the smaller of the two, and this is evident for the 100-m square blocks.

A reasonable maximum for the kriging variance is perhaps 100 cm², equivalent to a standard error of 10 cm. For 40 × 40-m blocks this gives a grid spacing of approximately 40 m and about 75 m for 100 × 100-m squares.

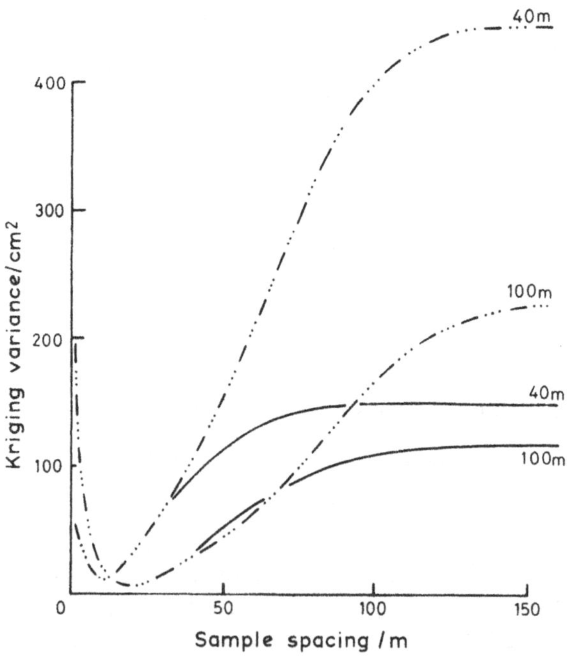


**Figure 35.** Graphs of maximum variance for kriging cover loam thickness over 40 × 40-m and 100 × 100-m blocks at Hole Farm. The solid lines are for square grids with the blocks centered over the grid nodes, the broken lines are for blocks centered in the centers of the grid squares.

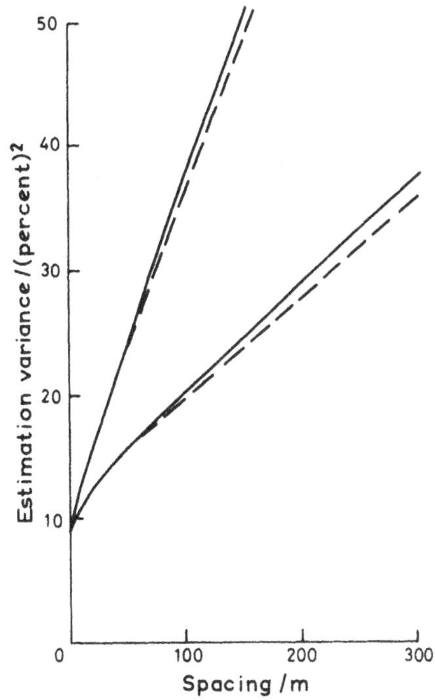


**Figure 36.** The relation between punctual kriging variance of stone content at Plas Gogerddan and sample spacing in the directions of maximum and minimum variation.

## A. Effect of Anisotropy

Equilateral sampling grids are best where variation is isotropic, and there are many regions in which this is so. As above, however, the soil can vary anisotropically, and in these situations some modification is needed. If the anisotropy is geometric the modification is simple.

The problem of determining the optimal intensity is treated initially as though variation were isotropic with the semi-variogram for the direction, $\phi$, in which the gradient or distance parameter is greatest. In the linear case this gives

$$\gamma(h) = c_0 + A|\mathbf{h}|, \qquad [93]$$

from equation [41]. The graph of $\sigma^2_{k_{\max}}$ against intensity is drawn and the optimal intensity read from it. This gives the spacing, $m$, in that direction. This can be repeated for direction $\phi + \pi/2$. Figure 36 shows such graphs for punctual estimation for the stone content in Cae Ruel whose semi-variogram is given earlier. However, the second graph is not needed since the spacing for this direction is easily computed by multiplying $m$ by the anisotropy ratio $A/B$.

Thus, the optimal strategy is to sample along parallel transects aligned in the direction of maximum variation. The spacing on the transects is $m$ and that between them $mA/B$. It is the familiar pattern of soil survey.

If costs are fixed then the best strategy is still as above. The sampling intervals are now such that the product $m \times mA/B$ equals the reciprocal of the sampling intensity.

## Acknowledgments

It is my pleasure to thank Dr T.M. Burgess and Dr A.B. McBratney whose collaboration produced so many of the results that I have quoted, and Mrs. Joyce Munden for much of the fresh computing and graphics. I also thank Dr P.A. Burrough for his advice on fractals and Dr H.E. Cuanalo for the Sandford data.

## References

Armstrong, M. 1984. Improving the estimation and modelling of the variogram. In: G. Verly, M. David, A.G. Journel, and A. Marechal (eds.), *Geostatistics for natural resource characterization*. Reidel, Dordrecht. pp 1–19.

Armstrong, M., and R. Jabin. 1981. Variogram models must be positive-definite. *Math. Geol.* 13:455–459.

Bartlett, M.S. 1966. *An introduction to stochastic processes with special reference to methods and applications*. 2nd ed. Cambridge University Press.

Beckett, P.H.T., and P.A. Burrough. 1971. The relation between the cost and utility in soil survey. V. The cost-effectiveness of different soil survey procedures. *J. Soil Sci.* 22:481–489.

Beckett, P.H.T., and R. Webster. 1971. Soil variability: a review. *Soils Fert.* 34:1–15.

Berry, M., and Z.V. Lewis, 1980. On the Weierstrass-Mandelbrot fractal function. *Proceedings of the Royal Society* A 370:459–484.

Burgess, T.M., and R. Webster. 1980a. Optimal interpolation and isarithmic mapping of soil properties. I. The semi-variogram and punctual kriging. *J. Soil Sci.* 31:315–331.

Burgess, T.M., and R. Webster. 1980b. Optimal interpolation and isarithmic mapping of soil properties. II. Block kriging. *J. Soil Sci.* 31:333–341.

Burgess, T.M., and R. Webster. 1984. Optimal sampling strategies for mapping soil types. I. Distribution of boundary spacings. *J. Soil Sci.* 35. 641–654.

Burgess, T.M., R. Webster, and A.B. McBratney. 1981. Optimal interpolation and isarithmic mapping of soil properties. IV. Sampling strategy. *J. Soil Sci.* 32:643–659.

Burrough, P.A. 1983a. Multiscale sources of spatial variation in soil. I. The application of fractal concepts to nested levels of soil variation. *J. Soil Sci.* 34:577–597.

Burrough, P.A. 1983b. Multiscale sources of spatial variation in soil. II. A non-Brownian fractal model and its application in soil survey. *J. Soil Sci.* 34:599–560.

Burrough, P.A. In press. The application of fractal ideas to geophysical phenomena. *Bull. Inst. Math. Appl.*

Campbell, J.B. 1978. Spatial variation of sand content and pH within single contiguous delineations of two soil mapping units. *Soil Sci. Soc. Am. J.* 42:460–464.

Chatfield, C. 1980. *The analysis of time series*. Chapman and Hall, London.

Christakos, G. 1984. On the problem of permissible covariance and variogram models. *Water Resour. Res.* 20:251–265.

Cochran, W.G. 1946. Relative accuracy of systematic and stratified random samples for a certain class of populations. *Ann. Math. Stat.* 17:164–177.

Cressie, N., and D.M. Hawkins. 1980. Robust estimation of the variogram. I. *Math. Geol.* 12:115–125.

Dalenius, T., J. Hajek, and S. Zubrzycki. 1961. On plane sampling and related geometrical problems. *Proc. 4th Berkeley Symp. Prob. Stat.* 1:125–150.

David, M. 1977. *Geostatistical ore reserve estimation*. Elsevier, Scientific Publishing Co., Amsterdam.

Delfiner, P. 1976. Linear estimation of non stationary spatial phenomena. In: M. Guarascio, M. David, and C. Huijbregts (eds.), *Advanced geostatistics in the mining industry*. Reidel, Dordrecht. pp. 49–68.

Dunn, M.R. 1983. A simple sufficient condition for a variogram model to yield positive definite variances under restrictions. *Math. Geol.* 15:553–564.

Gajem, Y.M., A.W. Warrick, and D.E. Myers. 1981. Spatial dependence of physical properties of a typic torrifluvent soil. *Soil Sci. Soc. Am. J.* 45:709–715.

Gandin, L.S. 1965. *Objective analysis of meteorological fields*. Israel Program for Scientific Translations, Jerusalem.

Harbaugh, J.W., and D.F. Merriam. 1968. *Computer applications in stratigraphic analysis*. John Wiley and Sons, New York.

Hawkins, D.M., and D.F. Merriam, 1974. Zonation of multivariate sequences of digitized geologic data. *Mathematical Geology* 6:263–269.

Jenkins, G.M., and D.G. Watts. 1968. *Spectral analysis and its applications*. Holden-Day, San Francisco.

Journel, A.G., and C.J. Huijbregts. 1978. *Mining geostatistics*. Academic Press, London.

Kendall, M.G., and A. Stuart. 1976. *The advanced theory of statistics. 3. Design and analysis, and time series*. 3rd ed. Griffin, London.

Kitanidis, P.K. 1983. Statistical estimation of polynominal generalized covariance functions and hydrological applications. *Water Resour. Res.* 19:909–921.

Kozlovskii, F.I., and N.P. Sorokina. 1976. The soil individual and elementary analysis of the soil-cover pattern. In: V.M. Fridland (ed.), *Soil combinations and their genesis*. Amerind Publishing Co., New Delhi. pp. 55–64. (Translated from Russian)

Krige, D.G. 1966. Two dimensional weighted moving average trend surfaces for ore-evaluation. *J. South African Inst. Min. Metal.* 66:13–38.

Mandelbrot, B.B. 1982. *The fractal geometry of nature*. W.H. Freeman, London.

Matérn, B. 1960. Spatial variation. Stochastic models and their application to some problems in forest surveys and other sampling investigations. *Meddelanden* från Statens Skogsforskningsinstitut 49:1–144.

Matheron, G. 1965. *Les variables régionalisées et leur estimation*. Masson, Paris.

Matheron, G. 1971. The theory of regionalized variables and its applications. *Cahiers du Centre de Morphologie Mathématique, Fontainebleau*, No. 5.

McBratney, A.B., and R. Webster. 1981a. Spatial dependence and classification of the soil along a transect in northeast Scotland. *Geoderma* 26:63–82.

McBratney, A.B., and R. Webster. 1981b. The design of optimal sampling schemes for local estimation and mapping of regionalized variables. II. Program and examples. *Comp. Geosci.* 7:335–365.

McBratney, A.B., and R. Webster. 1983. Optimal interpolation and isarithmic mapping of soil properties. V. Co-regionalization and multiple sampling strategy. *J. Soil Sci.* 34:137–162.

McBratney, A.B., R. Webster, and T.M. Burgess. 1981. The design of optimal sampling schemes for local estimation and mapping of regionalized variables. I. Theory and method. *Comp. Geosci.* 7:331–334.

McBratney, A.B., R. Webster, R.G. McLaren, and R.B. Spiers. 1982. Regional variation of extractable copper and cobalt in the topsoil of south-east Scotland. *Agronomie* 2:969–982.

Miesch, A.T. 1975. Variograms and variance components in geochemistry and ore evaluation. *Geol. Soc. Am. Mem.* 142:333–340.

Morse, R.K., and T.H. Thorburn. 1961. Reliability of soil maps. *Proc. 5th Int. Conf. Soil Mech. Found. Eng.* 1:259–262.

Nortcliff, S. 1978. Soil variability and reconnaissance soil mapping: a statistical study in Norfolk. *J. Soil Sci.* 29:403–418.

Oliver, M.A. 1984. Soil variation in the Wyre Forest: its elucidation and measurement. Ph.D. Thesis, Birmingham University.

Olea, R.A. 1975. *Optimum mapping techniques using regionalized variable theory*. Series on Spatial Analysis No. 2. Kansas Geological Survey, Lawrence.

Olea, R.A. 1977. *Measuring spatial dependence with semi-variograms*. Series on Spatial Analysis No. 3. Kansas Geological Survey, Lawrence.

Orey, S. 1970. Gaussian sample functions and the Hausdorff dimension of level crossings. *Zeitschrift für Wahrschein-lichkeitstheorie und Verwandte Gebiete* 15:249–256.

Quenouille, M.H. 1949. Problems in plane sampling. *Ann. Math. Stat.* 20:355–375.

Ross, G.J.S. 1980. *MLP Maximum Likelihood Program*. Rothamsted Experimental Station, Harpenden.

Russo, D., and E. Bresler. 1982. Soil hydraulic properties as stochastic processes. II. Errors of estimates in a heterogeneous field. *Soil Sci. Soc. Am. J.* 46:20–26.

Sisson, J.B., and P.H. Wierenga. 1981. Spatial variability of steady-state infiltration rates as a stochastic process. *Soil Sci. Soc. Am. J.* 45:699–704.

Thornburn, T.H., and W.R. Larsen. 1959. A statistical study of soil sampling. *J. Soil Mech. Found. Div. Proc. Am. Soc. Civ. Eng.* 85, SMS, 1–13.

Vauclin, M., S.R. Vieira, G. Vachaud, and D.R. Neilsen. 1983. The use of cokriging with limited field soil observations. *Soil Sci. Soc. Am. J.* 47:175–184.

Vieira, S.R., D.R. Neilsen, and J.W. Biggar. 1981. Spatial variability of field-measured infiltration rate. *Soil Sci. Soc. Am. J.* 45:1040–1048.

Walker, P.H., G.F. Hall, and R. Protz. 1968. Soil trends and variability across selected landscapes in Iowa. *Soil Sci. Soc. Am. Proc.* 32:97–101.

Webster, R. 1973. Automatic soil-boundary location from transect data. *Math. Geol.* 5:27–37.

Webster, R. 1977. Spectral analysis of gilgai soil. *Austr. J. Soil Res.* 15:191–204.

Webster, R. 1978. Optimally partitioning soil transects. *J. Soil Sci.* 29:388–402.

Webster, R. 1981. Experience of kriging from field measurements of soil properties. In: M.-C. Girard (ed.), *Traitement informatique des données de sol*. Tome 1. Institut National Agronomique, Paris-Grignon. pp. 101–109.

Webster, R., and P.H.T. Beckett. 1968. Quality and usefulness of soil maps. *Nature* (London) 219:680–682.

Webster, R., and T.M. Burgess. 1980. Optimal interpolation and isarithmic mapping of soil properties. III. Changing drift and universal kriging. *J. Soil Sci.* 31:505–524.

Webster, R., and T.M. Burgess. 1984a. Une approche probabiliste de la cartographie du sol. *Sciences de la Terre, Série Informatique Géologique* 18:175–185.

Webster, R., and T.M. Burgess. 1984b. Sampling and bulking strategies for estimating soil properties of small regions. *J. Soil Sci.* 35:127–140.

Webster, R., and B.E. Butler. 1976. Soil classification and survey studies at Ginninderra. *Austr. J. Soil Res.* 14:1–24.

Webster, R., and H.E. Cuanalo de la C. 1975. Soil transect correlograms of north Oxfordshire and their interpretation. *J. Soil Sci.* 26:176–194.

Webster, R., and S. Nortcliff. 1984. Improved estimation of micro nutrients in hectare plots of the Sonning series. *J. Soil Sci.* 35:667–672.

Xu Jiyan, and R. Webster. 1984. A geostatistical study of topsoil properties in Zhangwu County, China. *Catena* 11:13–26.

Yates, F. 1948. Systematic sampling. *Phil. Trans. Roy. Soc. Lond.* A 241:345–377.

Yost, R.S., G. Uehara, and R.L. Fox. 1982. Geostatistical analysis of soil chemical properties of large land areas. I. Semi-variograms. *Soil Sci. Soc. Am. J.* 46:1028–1032.

Youden, W.J., and A. Mehlich. 1937. Selection of efficient methods for soil sampling. *Contr. Boyce Thompson Inst. Plant Res.* 9:59–70.

Zubrzycki, S. 1957. On estimating gangue parameters. *Zastosowania Matematakyi* 3:105–153. (In Polish)

# The Effect of Green Manuring on the Physical Properties of Temperate-Area Soils

R.J. MacRae* and G.R. Mehuys*



## I. Introduction

Green manuring has received little attention by the research community in the past few decades. But with renewed concern about our soil resources, particularly soil degradation, green manuring is increasingly appealing as a means to reduce losses of soil organic matter, compaction, and soil erosion and still maintain economic returns. Recent attempts to reevaluate the usefulness of green manures have been hindered by a lack of readily

*Department of Renewable Resources, Macdonald College of McGill University, Ste. Anne de Bellevue, Quebec, Canada H9X 1CO


Advances in Soil Science, Volume 3

available and current information. This paper attempts to compile and synthesize what information is presently available.

The literature on green manuring on a global scale is voluminous but confusing. Joffe (1955) has explained the variability in green-manure research results with a zonality principle: The effectiveness of green manures is dependent on the climatic zone in which they are used. Literature cited in this review has been confined to the temperate regions as much as possible because the effects of green manures on soils of the humid and tropical areas are significantly different. Also, we have concentrated on studies involving field cropping systems, particularly those involving corn (*Zea mays*) and green manures, because corn production systems are so widespread in the temperate region and are one of the monocultures most detrimental to the soil physical condition.

Green manuring is the process of turning a crop into the soil, whether originally intended or not, irrespective of its state of maturity, for the purpose of effecting some agronomic improvement. It is an ancient practice. The earliest references to green manuring date from the Han dynasty of China, prior to 1134 B.C. Lupines and faba beans (*Vicia faba*) were used by the Greeks as green-manure crops before the third century B.C., and the practice has continued in southern Europe. The adoption of green manuring occurred much later in northern Europe (Pieters, 1927). In North America, green manuring has existed as a farming practice since the eighteenth century. Its popularity reached a peak in the first decades of this century at which time its practice was very widespread. The advent of cheap inorganic fertilizers resulted in its decline.

It is well known that there is a general relationship between soil physical properties and the amount of organic matter a soil contains. The general statement would be that, all other factors being equal, a soil with a high soil organic-matter level has a good physical condition. "Soil organic matter refers to the organic fraction of the soil; it includes plant and animal residues at various stages of decomposition, cells (living and dead) and tissues of microbes, and substances synthesized by the soil population" (Canada Department of Agriculture, 1972). Seldom does the addition or synthesis of such material not affect at least one soil parameter (Allison, 1973). If a green manure is capable of increasing the soil organic-matter level, then one could expect that the physical condition would be improved.

The effect of organic matter on soil physical parameters will be examined first, followed by a review of the effect of green manures on soil organic-matter levels. Because green-manure studies have usually been performed independently of the general examination of how organic matter affects physical properties, research performed on the effects of green manures on various soil physical parameters will be reviewed to determine if there is consistency with the results obtained in organic-matter studies. Finally, if green manures bring about some improvement in the soil physical condition, the question of how this may translate into improved crop performance will be discussed.

## II. Effect of Organic Matter on Soil Physical Properties

The properties most likely to be affected by increased organic-matter levels are aggregate distribution and stability, bulk density, moisture retention, and water movement, but the extent to which these parameters might be improved, and the conditions under which the improvement might occur, are not well defined.

### A. Aggregate Distribution and Stability

Martin *et al.* (1955) defined a soil aggregate as a "naturally occurring cluster or group of soil particles in which the forces holding the particle together are much stronger than the forces between adjacent aggregates." Aggregate formation and stabilization result from separate forces but often occur simultaneously. Few agents are involved in both processes.

Organic matter plays a minimal role in aggregate formation. Only in sandy soils, and perhaps clays, does it have any effect. Decomposed fungi mycelia may form aggregates in sandy soils through physical entanglement, particularly after the addition of plant residues, because fungi are closely associated with freshly incorporated plant tissue (Kononova, 1966). Hartmann and De Boodt (1974) found that sands of medium to high organic-matter content in The Netherlands exhibited an aggregating pattern similar to a sandy loam soil. They thus attributed to organic matter a major role in aggregate formation in coarse-textured soils. In fine-textured soils, coarse organic material may prevent the formation of large aggregates by blocking the attractive forces between smaller aggregates. But in general, agricultural soils have sufficient opportunity for aggregate formation and insufficient stabilizing material. It is in this area that organic matter plays an important role.

Organic matter is the primary stabilizing agent for aggregates in temperate soils. Specifically, the decomposition products of plant and animal residues are responsible for aggregate stabilization. These include microbial gums and mucilages, low molecular weight fulvic-acid molecules, and linear organic polymers like fats and waxes (Greenland *et al.*, 1962; Harris *et al.*, 1966; Allison, 1973). Theories of stabilization mechanisms are numerous. The most widely accepted view is that of Emerson (1959), who described organic binding at a microscopic level. In his model (Figure 1), organic material of molecular thickness acts as a binding agent between clay micelles or clay and quartzitic particles. Infrequently, the organic material coats clay domains. A series of linkages between clay and

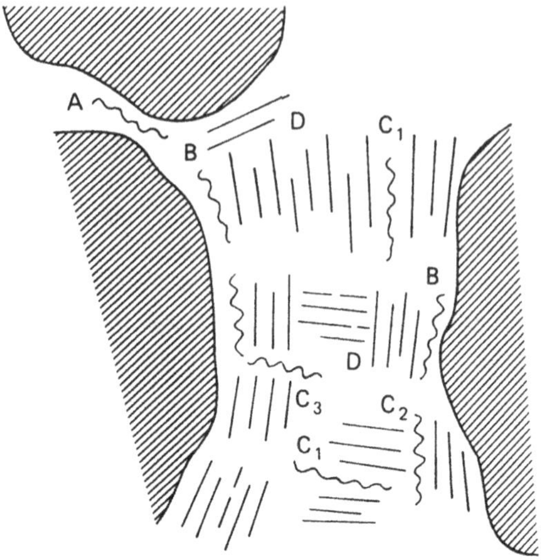


**Figure 1.** Possible arrangements of quartz particles, clay domains, and organic matter in a soil aggregate according to Emerson's model. (A) Quartz–organic colloid–quartz; (B) quartz–organic colloid–clay domain; (C) clay domain–organic colloid–clay domain, face to face ($C_1$), edge to face ($C_2$), edge to edge ($C_3$); (D) clay–domain edge–clay-domain face. The quartz particles can be linked by organic matter (A) or through two or more domains, which are themselves linked by organic matter (C). Some clay domains may be held together by electrostatic attraction between the positive edges and negative faces (D).

quartzitic material results in a soil crumb. This model has not, as yet, been experimentally proven.

Although Emerson's model is applicable over a wide range of textural classes, it may not completely explain aggregate stabilization in very sandy soils or clay soils. Soils with <10% clay generally act as if non-aggregated (Allison, 1973), and Emerson's model seems to support this. Pringle and Coutts (1956), however, found stable aggregates in a Scottish soil having 8.9% clay content and a loamy sand textural classification. In fine-textured soils, improving aggregation often involves the breakdown of large aggregates. In addition to the stabilization mechanisms that Emerson described, large, slightly decomposed organic molecules may be necessary to prevent slumping or the creation of a very finely divided system of aggregates (Greenland, 1965).

The highest percentage of stable aggregates is found in soils with high organic-matter content (Doyle and Hamlyn, 1960; Harris *et al.*, 1966). Although this has been proven chemically, it is sometimes difficult to obtain this result in agronomic studies. Greenland (1965) has indicated that organic carbon determinations may not correlate well with aggregate analyses because the total organic-carbon analysis commonly used in agronomic studies is not measuring just those organic materials responsible for aggregate stabilization. Chester *et al.* (1957) correlated soil gum and total organic-matter levels of silt loam soils in Wisconsin with aggregate stability and found only a slightly significant correlation with total soil organic matter, while soil gums were highly positively correlated to stability. De Kimpe *et al.* (1982) found no significant correlations between soil organic-matter levels and dry- and wet-sieve analyses in their study of 21 Québec topsoils. Coote and Ramsey (1983) did not find a significant correlation between organic carbon and wet mean weight diameter (MWD) in their study of four Ontario soils. These results could also be due to the inadequacies of sieving procedures (Harris *et al.*, 1966).

In summary, organic matter is known to play a primary role in aggregate stabilization, although agronomic studies have not been able to suitably quantify the relationship. Organic matter plays a minor role in aggregate formation.

### B. Bulk Density

With few exceptions, organic matter decreases the bulk density of soil. This may occur directly by dilution of the soil matrix with a less dense material, or indirectly by improving aggregate stability. This effect does not seem to be limited to certain soil textural classes. It is the indirect improvement of bulk density, through the improvement of aggregate distribution and stability, that is most important. Physical incorporation alone can affect detrimentally other physical parameters such as moisture movement (Jamison, 1960).

In a recent Québec study of 21 topsoils with organic-matter contents ranging from 1.6 to 11.9%, De Kimpe *et al.* (1982) found high negative correlations between organic matter and maximum bulk density after compaction. Millette *et al.* (1980), comparing cultivated (for 60 yrs) and forested Orthic Podzols, found a highly significant negative correlation between organic matter and bulk density. Soane (1975) found organic-matter levels to be highly negatively correlated with bulk density in compaction studies. The correlation of texture with bulk density was not significant. Coote and Ramsey (1983), working with four Ontario soils of different textural class, found organic matter to be highly negatively correlated with bulk density in tillage studies.

### C. Moisture Content and Availability

Organic matter increases available water in soils under specific circumstances only. It is the exception rather than the rule.

Peele *et al.* (1948) found significant correlations between organic-matter content and available water in sandy loam soils of South Carolina. Using their data, Jamison (1953) refined their conclusions. He found the same significant correlations between organic matter and available water in soils with less than 15% clay. With more than 15% clay, other factors were dominant in determining available water. Jamison and Kroth (1958) supported this finding with results from Mississippi soils. They found organic matter to influence available water only in soils of medium-low clay content (13 to 20%). Because they found available water to increase with silt content, they suggested that the organic matter tends to help form stable micro-aggregates of coarse-silt size. Their conclusion was that coarse silt had the greatest influence on available water. Reeve *et al.* (1973) also found significant correlations between coarse silt and available water. They felt that organic-carbon levels were not a primary factor in determining available water in mineral soils.

Although Singh (1962) claimed that organic material has a very high water retention capacity, Jamison (1953) and Feustal and Byers (1936) have concluded that most of the moisture is held at potentials below −15 bars, thus unavailable for plant growth. Feustal and Byers (1936) found nothing to be gained by adding peat moss to a clay loam soil (Table 1). Jamison (1953) found that a heavy addition of peat to a silty clay loam improved available water only slightly. They suggested that only in very sandy soils, with low available water, might organic-matter additions increase available water, though the volume of material necessary to effect such a result might be too large to be economically feasible. Stevenson (1974) supported this conclusion below −1 bar matric potential on a loamy sand when moisture content was determined volumetrically. He found no effect of adding peat to a sandy loam and a detrimental decline of available water in a silt loam. The minimum required addition of peat was

**Table 1.** Available Water in Soil and Soil–Peat Mixtures

| | Available water ($m^3$/100 $m^3$ material) | | | | |
|---|---|---|---|---|---|
| | Peat (% by volume) | | | | |
| Soil texture | 0 | 10 | 20 | 30 | 50 |
| Clay loam[a] | 14.3 | – | 11.5 | – | 9.5 |
| Silty clay loam[b] | 12.0 | – | – | – | 13.0 |
| Silt loam[c] | 27.7 | 23.2 | – | 22.8 | 21.7 |
| Sandy loam[c] | 5.8 | 6.7 | – | 6.2 | 8.2 |
| Loamy fine sand[a] | 6.0 | – | 6.5 | – | 7.2 |
| Loamy sand[c] | 3.3 | 3.7 | – | 4.4 | 5.1 |
| Quartz sand[a] | 1.2 | – | 4.3 | – | 5.6 |

[a]Calculated from Feustal and Byers (1936).
[b]Adapted from Jamison (1953). Peat incorporated in plow layer at the rate of 103 t/ha 5 yrs previously and calculated to be approximately 50% by volume.
[c]From Stevenson (1974).

substantial: 10% by volume of soil used (Table 1). Hartmann and De Boodt (1974), on the other hand, found that organic matter had no effect on the water retention properties (or critical capillary depression in their study) of sandy soils with organic-matter levels ranging from 0.12 to 7.17%.

De Kimpe *et al.* (1982) found a positive correlation between organic matter and gravimetric water content at field capacity on a wide range of Québec soils. There was also a significant negative correlation between bulk density and organic carbon. Stevenson (1974) has shown that changes in gravimetric water content with organic matter are often not a function of improvements in moisture retention but rather a function of the lower bulk density (Table 2) and that moisture content has to be measured volumetrically to really evaluate the relationship. Thus, the correlation between organic matter and gravimetric water content in the De Kimpe *et al.* (1982) study is likely a function of the relationship between bulk density and organic matter. Confirming the data in Table 2, MacRae[1] found a significant correlation between organic carbon and volumetric moisture retention on a loamy sand. There was no significant correlation between organic carbon and volumetric or gravimetric moisture retention of a clay soil used in the study.

Improvements in other physical properties induced by organic-matter additions sometimes increase water retention. Such is not the case for aggregation. Jamison (1953) found increased aggregation in three soils of various textural classes to result in decreased available water. This result was due, in part, to an increase in moisture retention at the permanent wilting point (−15 bar). Doyle and Hamlyn (1960) found similar results. Hillel (1980) showed that increased aggregation can decrease the micropore volume and decrease the water content at field capacity, thus decreasing available water. Jamison (1953) has stated that only in poorly drained clayey soils does increased aggregation improve available water. He was of the opinion that the increased macropore space is part of the defined available water in poorly drained soils. He felt that plant responses to improved aggregation in soils were due to the increased rooting volume from which water could be taken and not to improvements in water retention in a given volume. Tamboli *et al.* (1964) have found that aggregates with the greatest organic-matter content do not have the greatest moisture retentive ability. In bulk-density studies conducted in England and Wales, Reeve *et al.* (1973) found that bulk density correlated negatively with available water in A horizons of silty soils only. B and C horizons gave negative correlations in all textural classes; however, not all correlations in this study were significant.

---

[1]Master of Science Thesis, McGill University, 1983.

**Table 2.** Soil Water Retention as a Function of Soil Texture and Added Organic Matter[a]

| Soil texture | Peat moss added (% by vol.) | Bulk density ($kg/m^3$) | Water retained between −0.3 and −15.0 bars matric potential: Mass basis (%) | Volume basis measured (%) | Volume basis calculated[b] (%) |
|---|---|---|---|---|---|
| Loamy sand | 0 | 1460 | 2.2 | 3.3 | 3.2 |
| | 10 | 1410 | 2.5 | 3.7 | 3.5 |
| | 30 | 1310 | 3.4 | 4.4 | 4.5 |
| | 50 | 1210 | 4.1 | 5.1 | 4.9 |
| Sandy loam | 0 | 1200 | 4.9 | 5.8 | 5.9 |
| | 10 | 1170 | 5.6 | 6.7 | 6.6 |
| | 30 | 1090 | 5.6 | 6.2 | 6.1 |
| | 50 | 1020 | 8.0 | 8.2 | 8.2 |
| Silt loam | 0 | 1160 | 23.8 | 27.7 | 27.6 |
| | 10 | 1110 | 20.9 | 23.2 | 23.2 |
| | 30 | 1020 | 22.3 | 22.8 | 22.7 |
| | 50 | 960 | 22.3 | 21.7 | 21.4 |

[a]Calculated from Stevenson (1974).
[b]Calculated as percent dry mass × bulk density/density of water.

### D. Water Movement

Organic matter has a strong effect on infiltration of water into soils. Wischmeier and Mannering (1965), working with 44 Corn-Belt soils with organic-matter contents ranging from 1 to 14%, found that, of all variables examined, organic matter was the most closely correlated with increases in infiltration. Textural differences had little influence. The reason for this relationship is somewhat unclear.

Allison (1973) stated that, when well mixed in the soil, organic matter greatly aids infiltration by providing better aggregation and structure, consequently lowering bulk density and easing water movement. Martin and Richards (1969) found, however, that increases in aggregate stability did not necessarily result in increased hydraulic conductivity in a sandy loam aggregated by various bacterial polysaccharides. In fact, improvement in hydraulic conductivity was noted in some cases without an increase in aggregate stability. Wischmeier and Mannering (1965) obtained similar results: Their aggregation index for wet aggregates was poorly correlated with infiltration ($r^2 < 0.10$) even though organic matter was highly correlated with infiltration.

In some sandy soils, decreases in infiltration are desirable, and organic matter can play an important role in this also. Working with a synthetic sandy loam, Barley (1954) found coarse organic material to decrease permeability. These results confirmed those of a previous study (Barley, 1953) in which permeability decreased as the coarse organic-matter content of a sandy loam soil increased in the surface 7.5 cm from 0.6 to 2.2%.

De Kimpe *et al.* (1982) found that organic matter had little, if any, effect on the saturated hydraulic conductivity of soils in a state of maximum organization or maximum compaction. Their study did not determine the relationship between organic matter and saturated hydraulic conductivity for spatial arrangements normally found in the field as did the studies mentioned above.

## III. Effect of Green Manures on Soil Organic-Matter Levels

It is generally accepted that green manures will maintain or increase organic matter or maintain or increase soil nitrogen levels but not both at the same time (Allison, 1973; Warman, 1980). Joffe (1955) stated that green manures were originally used as fertilizer; they were not perceived as a means to increase the organic-matter content of the soil until the late nineteenth century when the trend towards a specialized and mechanized agriculture began. Pieters (1927) felt that the most important function of green manuring was to increase soil organic-matter levels, though he later modified this, concluding "that in the main the object of green manuring must be to maintain rather than increase the quantity of organic matter in soils" (Pieters and McKee, 1938). Some have not accepted that green manures can even maintain organic-matter levels (Pinck *et al.*, 1946; Poyser *et al.*, 1957). Russell (1973) stated that on a global level green manure has been much more effective as a nitrogen source than as a source of organic matter. Most researchers at least concede that green manures will affect the rate of loss of organic carbon in cropping systems and its ultimate level in the soil.

The maintenance or accumulation of organic matter in soil is dependent on a number of factors such as the chemical nature of the added material, soil and climatic factors as they affect microbial activity, and cultural practices (Table 3). The popular conclusion is that a plant material resistant to ready decomposition is necessary if soil organic-matter levels are to be maintained or increased. Warman (1980) stated that plant material typically low in nitrogen, i.e., 1.5% nitrogen or less on a dry-weight basis, can be effective in improving the organic-matter content of a soil. From the literature on leguminous and non-leguminous green manures, however, no clear-cut relationship emerges with regard to organic-matter accumulation and nitrogen content in the soil (Table 4). Prince *et al.* (1941), in a 40-yr potting study, found that vetch (*Vicia sativa*), incorporated 2 yrs in a 5-yr rotation of corn–oats (*Avena sativa*)–oats–wheat (*Triticum vulgare*)–timothy (*Phleum pratense*), did not maintain the organic-matter content at its original level, unless farm manure was also added and the soil limed. Poyser *et al.* (1957), working with clay soils in Manitoba, found the organic-carbon levels in the soil to decrease across all treatments over a 25-yr period. Losses of organic carbon in a green manure–wheat–corn–wheat rotation ranged from 30.7% when the green manure was sweet clover to 6.1% when it was peas. Losses from nonleguminous green-manure treatments were slightly greater, ranging from 37.8 to 26.3%. Chater and Gasser (1970), working on the Woburn long-term green-manuring experiment in England, found trefoil (*Lotus corniculatus*), incorporated each year between 1936 and 1965, to increase organic-matter levels slightly. With yearly incorporation of ryegrass (*Lolium perenne*) over the same period, soil organic-matter levels declined slightly, even though more organic material and less nitrogen were added over the course of the experiment. They concluded that, on the whole, yearly incorporation of green manures could maintain original levels of soil organic matter. Mann (1959), reporting on the first 18 yrs of the Woburn experiment, found that the manurial value was not directly connected with the nitrogen levels of the plant tissue nor with the total amount of organic material buried. Allison (1973) reached the same conclusion with regard to amount of material buried but disagreed with Mann's conclusions on tissue nitrogen levels. De Haan (1977), working with a sandy and a clayey soil in The Netherlands, found no correlation between the percent nitrogen of the added tissue and the amount of organic matter accumulated over 10 yrs, though he did find highly significant correlations with the percent lignin of the original material. Leuken *et al.* (1962), working with Saskatchewan

**Table 3.** Factors Influencing the Maintenance and Accumulation of Organic Matter in Soil

| | |
|---|---|
| *Climate* | |
| | Temperature, solar radiation |
| | Precipitation, evaporation |
| *Soil/Site* | |
| | Elevation, slope, aspect, geographical location |
| | Soil type |
| | Texture |
| | Structure, compaction |
| | Native organic matter and humus content |
| | Soil temperature |
| | Soil moisture, aeration |
| | pH |
| | Mineral ion content |
| | Plant cover (species, density, distribution, history of site) |
| | Microbial populations (species, density, distribution, history of site) |
| | Faunal populations (species, density, distribution, history of site) |
| | Use of fertilizers, lime, mulches, and pesticides |
| | Tillage, cultivation, drainage, irrigation |
| | Fire, e.g., burning of crop residues |
| *Material Incorporated* | |
| | Composition (e.g., carbohydrates, proteins, lignins, fats, waxes) |
| | Quantity added per unit area |
| | Moisture content |
| | C:N ratio |
| | Mineral ion content |
| | Timing, method, and frequency of incorporation |
| *Experimental Variables* | |
| | Plot size, arrangement, number of replicates |
| | Cultivation practices |
| | Frequency of sampling |
| | Sample size, shape |
| | Techniques for measuring variables |

clayey soils, concluded that the nitrogen content of incorporated plant material was very important in determining the degree of organic-matter accumulation.

These differing results are likely a function of other factors affecting organic-matter accumulation. Any soil factor that is likely to affect the soil microbial and faunal populations is likely to have an effect on organic-matter maintenance or accumulation. These include such factors as soil

**Table 4.** Effect of Long-Term Green Manuring on Organic Matter and Nitrogen Content of Soils

| Study | Length of study (yrs) | Green manure used | Rotation (X=green manure) | Soil type | Initial soil OM (%) | Initial soil N (%) | Changes at end of study: Organic matter | Changes at end of study: Soil N |
|---|---|---|---|---|---|---|---|---|
| *Potting studies* | | | | | | | | |
| Prince *et al.* (1941) | 40 | Vetch | Corn–X–oats–X–oats–wheat–timothy | Loam | 3.9 | 0.197 | Decrease | Decrease |
| Leuken *et al.* (1962) | 150-day incubation | Legumes and non-legumes | No crop | Silty clay | 10.3 | 0.491 | Increase | N.R.[a] |
| | | | | Clay loam | 2.8 | 0.131 | Increase | N.R. |
| De Haan (1977) | 10 | Legumes and non-legumes | No crop, annual incorporation | Sand | 3.9 | 0.156 | Increase | Increase |
| | | | | Clay | 3.0 | 0.224 | Increase | Increase |
| *Field studies* | | | | | | | | |
| Poyser *et al.* (1957) | 25 | Legumes and non-legumes | X–wheat–corn–wheat | Clay | 7.8 | 0.370 | Decrease | Decrease |
| Mann (1959) | 18 | Legumes and non-legumes | X–kale[b] or cabbage–barley | Sandy loam | 1.48 | 0.090 | Increase | Increase |
| Chater and Gasser (1970) | 30 | Trefoil | X–barley[c]–potatoes[d] | Sandy loam | 1.48 | 0.090 | Same | Increase |
| [cont. of Mann (1959)] | | Ryegrass | X–barley | Sandy loam | 1.48 | 0.090 | Same | Decrease |

[a]N.R. = not reported. [b]*Brassica oleracea.* [c]*Hordeum vulgare.* [d]*Solanum tuberosum.*

type, soil nitrogen levels, native soil organic-matter levels—although no trend can be detected with respect to these in Table 4—soil moisture, temperature, and exchangeable cations (Table 3). The following study provides an example. Sowden and Atkinson (1968) applied annual additions of the same leguminous and non-leguminous green manures to Uplands sand and Rideau clay. Besides textural differences, the soils differed in pH, initial organic-carbon level, and initial soil nitrogen (Table 5). After 20 yrs, they observed on the clay soil a loss of organic matter on the control and rye (*Secale cereale*) green-manure plots. All other treatments maintained soil organic-matter levels. On the sand, only the control plot lost organic matter; all other treatments increased organic-matter levels.

The authors did not try to explain which factors and their interactions might have been responsible for the different responses, but De Haan (1977) concluded that clay content was not a significant factor in determining the amount of organic matter formed. pH was a highly significant factor in his studies. In the Sowden and Atkinson (1968) study, the lower pH of the clay soil likely slowed the rate of organic-matter decomposition. Normally, organic matter would then accumulate. Because it did not, other factors were of greater influence.

Several authors have postulated that, in arable soils, an equilibrium soil organic-matter level exists (Jenny, 1930; Joffe, 1955; Allison, 1973). If this is so, the initial soil organic-matter level may affect the rate or amount of accumulation of new organic material. According to this theory, the potential of cultivation to reduce organic-matter levels is much greater than that of any cultural practice that attempts to increase the organic-matter level of the soil. But once a cultivated soil reaches its equilibrium level, no management practice can lower it, and it would then be more feasible to

**Table 5.** Soil Conditions Prior to and After a 20-yr Study on Annual Additions of Organic Materials[a]

| | | Before experiment | | | After 20 yrs |
|---|---|---|---|---|---|
| Soil series | Treatment | pH | Soil nitrogen (%) | Organic carbon (%) | Organic carbon (%) |
| Rideau clay | Control | 5.6 | 0.238 | 2.91 | 2.44 |
| | Green manures (mean) | 5.5 | 0.240 | 2.93 | 2.80 |
| Uplands sand | Control | 6.7 | 0.129 | 1.37 | 1.23 |
| | Green manures (mean) | 6.7 | 0.116 | 1.20 | 1.51 |

[a]From Sowden (1968) and Sowden and Atkinson (1968).

increase soil organic-matter levels by incorporating plant material (Joffe, 1955). It is possible that the sandy soil in the study of Sowden and Atkinson (1968) was closer to this equilibrium level than the clay soil. If a soil has not reached that equilibrum level, organic-matter additions may be able to retard the rate of decline or may, as some have suggested (Broadbent and Norman, 1947), increase the decomposition of the native organic matter, a process known as "priming." Priming would actually increase the rate of decline. Allison (1973), however, reviewed the literature on this effect and concluded that it is insignificant in the course of a growing season.

Low soil nitrogen levels will affect the rate of decomposition and the amount of material that decomposes, depending of course on other soil and climatic factors and particularly the nitrogen content of the material being added. Generally, low soil nitrogen levels will result in a slower decomposition of organic material but, over the long term, there is potential for greater accumulation. The work done by Sowden and Atkinson (1968) demonstrates this result (Table 5).

Green manures will affect soil organic-matter levels, but many factors play a role in determining what that effect will be. There is agreement on what the major factors are but not necessarily on which factors exert the most influence, of what magnitude the effects will be, and how they interact. The conclusions of Allison (1973) and Warman (1980), outlined at the beginning of this section, are appropriate only when all these factors are controlled.

## IV. Effects of Green Manures on Soil Physical Properties

### A. Aggregate Distribution and Stability

Although Ram and Zwerman (1960) have reported that aggregate stability "is positively correlated with the quality and quantity of organic material added to the soil," an examination of their citations and other work reveals that effects and effectiveness vary with soil texture and green manure used and that effects do not last long unless continuous additions are made.

Silt loam soils seem to be the most responsive to aggregate-stability increases from green manuring (Browning and Milam, 1944; Chester *et al.*, 1957; Wisniewski *et al.*, 1958; Ram and Zwerman, 1960). Effects are more variable in clayey and sandy soils with seemingly little association between increased aggregate stabilization and green manure used, frequency of incorporation, or number of years incorporated. Browning and Milam (1944) found increased aggregate stability in a silty clay loam but suggested that effects would be minimal in clays and in sandy soils. Mortensen and Young (1960), working with a silty clay in Ohio, found only transient improvements in aggregate stability attributable to the action of grass green manures. The clovers used in the study had no significant effect. Guttay *et al.* (1956) found no improvement in aggregate distribution or stability in a Michigan loam soil from biannual additions of sweet clover (*Melilotus alba*). Bowren and Wilson (1959), on the other hand, working with a Saskatchewan fine sandy loam, found a 3-yr rotation of sweet clover–wheat–oats to have larger dry aggregates than a fallow–wheat–oats rotation and more water-stable aggregates in the top 15 cm (Bowren and McNaughton, 1967).

Work done with a sandy loam in New Jersey by Benoit *et al.* (1962) indicates that several years of annual additions may be necessary before aggregate stability increases (3 yrs in their study). They used ryegrass and found that incorporation of tops alone did not improve aggregate stability. It is not clear from their study whether the structural improvements caused by green manuring occurred continuously over a 3-yr period. The effects may have been transitory. Halstead and Sowden (1968) did not find ryegrass to increase the aggregate stability of a Rideau clay after 20 yrs of annual additions. Alfalfa (*Medicago sativa*) as a green manure increased the aggregate stability somewhat, and an increase in organic carbon was associated with this improvement. The strength of the association was not determined. Rennie *et al.* (1954) found that more readily decomposable, incorporated material (legumes) produced more stable aggregates in the very short term in a laboratory study, but that non-leguminous plant material produced effects that were more persistent. That all the field studies mentioned above did not reach the same conclusion is likely because of the multitude of factors (Table 3) that affect decomposition of organic matter.

### B. Bulk Density

The addition of organic materials to a soil zone will automatically decrease the bulk density of that zone because the added material is of lower density than the soil matrix. The type of material used to effect such a result is largely inconsequential. Morachan *et al.* (1972) found no difference between alfalfa and corn stalks used to lower the bulk density of a silty clay loam subsoil. They felt the result might have been due "partly or totally to the physical incorporation of the chopped residue." Their study did not take place over a sufficient time period to evaluate the effects of the respective decomposition products.

The relationship between an improved aggregate condition and a reduction in soil bulk density is theoretically straightforward, but agronomic results do not consistently confirm this. Mortensen and Young (1960) did not find any change in the bulk density of a silty clay in the first growing season owing to addition of various green manures even though the aggregate stability of the soil was higher. In a companion study, there were bulk-density effects owing to green-manure treatments that were not simply a function of dilution (alfalfa and red clover [*Trifolium pratense*] were more effective in reducing bulk density than yellow sweet clover [*Melilotus officinalis*], ryegrass, and orchard grass [*Dactylis glomerata*]), but there were no effects on aggregate stability. Ram and Zwerman (1960) did find differences in aggregate stability associated with bulk density changes, as did Benoit *et al.* (1962).

There are also differences in the amount of time required to show effects and the duration of these effects (Table 6). As mentioned above, Morachan *et al.* (1972) found effects immediately after incorporation, but this was thought to be a purely mechanical effect. Mortensen and Young (1960) found 2 yrs had to pass before effects were apparent. Ram and Zwerman (1960) found that bulk density was lower during the first 2 yrs after incorporation but increased in the third. Benoit *et al.* (1962) found 3 successive yrs of ryegrass additions were required before a decrease in bulk density was apparent but it lasted for the following 3 yrs. De Haan (1977), in a potting study, found that all green manures used in his experiment, applied annually for 10 yrs, decreased bulk density, particularly in sandy soil. Bowren and McNaughton (1967) found that there were no differences between two 3-yr rotations after 28 yrs of cropping on fine sandy loam in Saskatchewan. The rotations differed 1 yr in 3 as one contained sweet clover as green manure, the other lay fallow. There were differences, however, in aggregate stability. Hageman and Shrader (1979) found no significant differences in bulk density in a 20-yr comparison of continuous corn and a corn–oats–meadow–meadow (as green manure) rotation. They attributed this result, in part, to the location of the sampling area in the

**Table 6.** Bulk Density Changes Associated with Incorporation of Green Manures

| Study | Duration of study (yrs) | Years after initiation of study | | | | | | |
|---|---|---|---|---|---|---|---|---|
| | | 1 | 2 | 3 | 4 | 5 | 6 | end |
| Mortensen and Young (1960) | 3 | 0[a] | 0 | – | | | | – |
| Ram and Zwerman (1960) | 7 | – | – | + | | | | – |
| Benoit *et al.* (1962) | 6 | 0 | 0 | 0 | – | – | – | – |
| Bowren and McNaughton (1967) | 28 | | | | | | | 0 |
| Morachan *et al.* (1972) | 13 | | | | | | | – |
| De Haan (1977) | 10 | – | – | – | – | – | – | – |
| Hageman and Shrader (1979) | 20 | 0 | 0 | 0 | 0 | 0 | 0 | 0 |

[a]0 = no change; + = increase; – = decrease.

continuous-corn plots. These areas did not receive any compaction, which suggests that degree of traffic and compaction may be major factors in bulk density studies. The studies mentioned above do not discuss the amount of compaction experimental units received and whether there were differences between treatments owing to the management system. One can speculate that more information on the experimental conditions such as those listed in Table 3 would provide more consistency in the published results and facilitate drawing definitive conclusions.

### C. Moisture Content and Availability

The literature is sparse regarding the relationship between green manures and soil moisture properties. What exists suggests that water retention may be improved in sandy soils. This is consistent with literature discussed earlier in connection with organic matter.

Singh (1962) and De Haan (1977) have suggested that the total moisture retention capacity of sandy soils will be increased by a variety of organic materials. De Haan (1977) did not find improvements in the clay soil he studied in a potting experiment. Singh (1962) stated that the organic material that results from green-manure additions is not colloidal in nature and therefore cannot improve water retention in most soils. This is likely true in the short term but in the long term all organic material added and decomposed becomes colloidal. It is possible that aerobic conditions in sandy soils result in more rapid decomposition than occurs in fine-textured soils (Allison, 1973), and this produces in a shorter time interval an organic material in the soil that has a higher capacity to retain water.

Morachan *et al.* (1972) found increased water retention in a silt loam soil with increased amount of residue added (the maximum addition was 16 t/ha). Ram and Zwerman (1960) attributed a higher soil moisture content to treatments with cover crops, but their study did not indicate whether this moisture increase was plant available. Although they had significant results on a gravimetric basis, they did not on a volumetric one. Morachan *et al.* (1972) did not do a volumetric determination. This might have changed their results and conclusions, as Jamison (1953) and Stevenson (1974) have indicated (Table 2).

### D. Water Movement

Most of the studies on the effects of green manures on water movement have been carried out on irrigated lands of the southern United States. Very few have been done on the more northerly temperate soils.

Several authors have reported favorable improvements in infiltration and hydraulic conductivity when aggregate stability was also improved (Browning and Milam, 1944; Benoit *et al.*, 1962). Browning and Milam (1944) concluded that increased stability in the macro-aggregate size range was of primary importance in effecting this result. Joffe (1955) concluded from his review of green manures in temperate soils that physical incorporation is also beneficial for increasing infiltration. Browning and Milam (1944) agreed. Jamison (1960), however, pointed out the limitations of physical incorporation. His study showed that isolation of soil zones can result in a decrease in moisture movement and air permeability that can cause yield decreases. Morachan *et al.* (1972) did not observe any improvement in infiltration caused by physical incorporation of various amounts of various residues.

The paucity of green-manure experimentation in temperate climates makes it difficult to conclude whether results are consistent with more general work on the organic-matter physical-property relationship. Few authors have properly examined the relationship between green manures, organic matter, and physical parameters. In general, though, there is some consistency with results discussed in the previous section.

## V. Effect of Green Manures on Crop Performance Related to Improvements in the Soil Physical Condition

Undoubtedly, improvement in the soil physical condition provides the opportunity for increased crop growth. This is an accepted fact. It is difficult, however, to assess quantitatively the degree of improvement associated with green manures and organic-matter levels necessary to effect a significant crop-yield increase, or which physical parameters have the most marked influence on crop production, or under what conditions yield increases will occur.

To use corn production as an example, most authors who have reported on the interaction of green manuring with corn yields have found increased corn yields associated with improvements in the physical condition but have been unable, because of experimental design, to differentiate the effect of improved fertility status from the improved physical environment (Guttay *et al.*, 1956; Bowren and McNaughton, 1967; Halstead and Sowden, 1968; Guernsey *et al.*, 1969; De Haan, 1977; Sheard, 1977).

Sopher and McCracken (1973) made a comprehensive attempt to determine the relationships between a variety of soil factors and corn yields. Using standard regression techniques, they attempted to determine cause-and-effect relationships between a variety of factors, including organic-matter levels and available water, and corn yields. Because of the high correlations between independent variables, however, they were unable to sufficiently define the factors of greatest importance.

Strickland (1951) and Morachan *et al.* (1972) found no correlations between physical improvement in silt loam soils and corn yield. They both concluded that the physical condition was not limiting to yields on the soils used. Page and Willard (1947) did find significant positive correlations between aggregate stability and corn yield in several clay soils of Ohio.

Strickland (1951) suggested that effects would be more apparent on clay soils than on silt soils. Dirks and Bolton (1980), working with a clay soil in Ontario, did find bulk density to have substantial indirect and direct effects on yield of grain corn. Compacted soil reduced nutrient uptake and root growth. Fertility factors also played a major role in determining yields.

Others have concluded that physical properties will only significantly affect corn yield in years of abnormal total moisture. De Boodt *et al.* (1953) found no significant correlations between corn yield and any physical property affected by green manuring with sweet clover in a corn–oats rotation. This result was attributed to a sufficient moisture supply in the year the correlations were determined. Yield differences were thought, thus, to be caused by fertility factors. Benoit *et al.* (1962) found significant correlations between corn yield and aggregate stability of a sandy loam only one year in four and attributed this result to adverse weather conditions in that year. Mortensen and Young (1960) found aggregate stability and aeration porosity to be positively correlated with the number of corn plants per hectare, ear count, and yield of corn in a dry year only. They concluded that the improved soil structure aided germination. Owing to the design of their study, their results are most likely a correlation of VAMA (vinyl acetate-maleic anhydride)-improved, rather than green manure-improved, aggregate stability and aeration porosity with corn yield. De Boodt *et al.* (1961) concluded that significant correlations of soil structure with crop yield are weather dependent.

## VI. Conclusions

The literature indicates that: (1) Organic matter does not directly affect all physical parameters. Aggregate stability and bulk density are the parameters most likely to be affected. Improvements in these parameters may bring about improvements in water movement or retention. (2) Green manures maintain soil organic-matter levels under particular, though not well-defined, soil conditions, and different plant species used as green manures can vary widely in their effect. Rarely will soil organic-matter levels be increased by green manuring. (3) Green manures do not necessarily improve the soil physical condition even when total soil organic-matter levels are maintained or increased by green-manure additions. Meaningful conclusions on this topic are difficult because of the paucity of information on the relationship between green manuring, soil organic-matter levels, and soil physical properties. (4) If the soil physical condition is improved, there may not be an associated improvement in crop performance. A growing season with abnormal weather conditions is the time when increased yields are most possible.

In reviewing the literature on the topic, the nature of the experiments performed has been a limiting factor in drawing conclusions. The number and complexity of factors that determine how green manuring will affect soil physical properties necessitates the use of non-traditional scientific approaches. It is not sufficient to look at a few parameters as many of the studies have done (and few have even examined the same set of parameters). For meaningful, general conclusions to be drawn, a holistic approach must be taken. In such an approach, many factors are measured and the interactions between them established by systematically holding one factor constant and measuring the effect upon the others. These effects must be measured in space and time, short and long term. It is not appropriate, as the traditional scientific approach dictates, to control all but the few factors to be studied. Such is partly the reason for the confusing body of information that exists in the published literature. The confusion will not likely disappear until a holistic approach is taken.

## References

Allison, F.E. 1973. *Soil organic matter and its role in crop production.* Elsevier Scientific Publishing Co., New York.

Barley, K.P. 1953. The root growth of irrigated perennial pastures and its effects on soil structure. *Austr. J. Agr. Res.* 4:283–291.

Barley, K.P. 1954. Effects of root growth and decay on the permeability of a synthetic sandy loam. *Soil Sci.* 78:205–209.

Benoit, R.E., N.A. Willits, and W.J. Hanna. 1962. Effect of rye winter cover crop on soil structure. *Agron. J.* 54:419–420.

Bowren, K.E., and W.N. McNaughton. 1967. The effect of controlled cropping on degraded soils. *Can. Dep. Agr. Publ.* No. 1321, Ottawa.

Bowren, K.E., and A.E. Wilson. 1959. Crop rotation studies on White Fox soils 1936–1957. *Can. Dep. Agr. Misc. Publ.*, Ottawa.

Broadbent, F.E., and A.G. Norman. 1947. Some factors affecting the availability of organic nitrogen in soil—a preliminary report. *Soil Sci. Soc. Am. Proc.* 11:264–267.

Browning, G.M., and F.M. Milam. 1944. Effect of different types of organic materials and lime on soil aggregation. *Soil Sci.* 57:91–106.

Canada Department of Agriculture. 1972. Glossary of terms in soil science. *Can. Dep. Agr. Publ.* No. 1459, Ottawa.

Chater, M., and J.K.R. Gasser. 1970. Effects of green manuring, farmyard manure and straw on the organic matter of soil and of green manuring on available nitrogen. *J. Soil Sci.* 21:127–137.

Chester, G., D.J. Attoe, and O.N. Allen. 1957. Soil aggregation in relation to various soil constituents. *Soil Sci. Soc. Am. Proc.* 21:272–277.

Coote, D.R., and J.F. Ramsey. 1983. Quantification of the effects of over 35 years of intensive cultivation on four soils. *Can. J. Soil Sci.* 63:1–14.

De Boodt, M.F., L. De Leenheer, and D. Kirkham. 1961. Soil aggregate stability indices and crop yields. *Soil Sci.* 91:138–146.

De Boodt, M.F., A.J. Englehorn, and D. Kirkham. 1953. Fall vs. spring plowing and soil physical conditions in a rotation experiment. *Agron. J.* 45:257–261.

De Haan, S. 1977. Humus, its formation, its relation with the mineral part of the soil, and its significance for soil productivity. In: *Soil organic matter studies. Vol. II.* Int. Atomic Energy Agency, Vienna. pp. 21–30

De Kimpe, C.R., M. Bernier-Cardou, and P. Jolicoeur. 1982. Compaction and settling of Québec soils in relation to their soil-water properties. *Can. J. Soil Sci.* 62:165–175.

Dirks, V.A., and E.F. Bolton. 1980. Regression analyses of grain yield of corn, level of leaf N P K and soil conditions in a long-time rotation experiment on Brookston clay. *Can. J. Soil Sci.* 60:599–611.

Doyle, J.J., and F.G. Hamlyn. 1960. Effects of different cropping systems and of a soil conditioner (VAMA) on some soil physical properties and on growth of tomatoes. *Can. J. Soil Sci.* 40:89–98.

Emerson, W.W. 1959. The structure of soil crumbs. *J. Soil Sci.* 10:235–244.

Feustal, I.C., and H.G. Byers. 1936. The comparative moisture-absorbing and moisture-retaining capacities of peat and soil mixtures. *USDA Tech. Bull.* No. 532.

Greenland, D.J. 1965. Interaction between clays and organic compounds in soils. II. Adsorption of soil organic compounds and its effects on soil properties. *Soils Fert.* 28:521–532.

Greenland, D.J., G.R. Lindstrom, and J.P. Quirk. 1962. Role of polysaccharides in stabilization of natural soil aggregates. *Nature* 191:1283–1284.

Guernsey, C.W., J.B. Fehrenbacher, B.W. Ray, and L.B. Miller. 1969. Corn yields, root volumes and soil changes on the Morrow Plots. *J. Soil Water Conserv.* 24:101–104.

Guttay, J.R., R.L. Cook, and A.E. Erickson. 1956. The effect of green and stable manure on the yield of crops and on the physical condition of a Tappan-Parkhill loam soil. *Soil Sci. Soc. Am. Proc.* 20:526–528.

Hageman, N.R., and W.D. Shrader. 1979. Effects of crop sequence and nitrogen fertilizer levels on soil bulk density. *Agron. J.* 71:1005–1008.

Halstead, R.L., and F.J. Sowden. 1968. Effect of long-term additions of organic matter on crop yields and soil properties. *Can. J. Soil Sci.* 48:341–348.

Harris, R.F., G. Chesters, and O.N. Allen. 1966. Dynamics of soil aggregation. *Adv. Agron.* 18:107–160.

Hartmann, R., and M. De Boodt. 1974. The influence of the moisture content, texture and organic matter on the aggregation of sandy and loamy soils. *Geoderma* 11:53–62.

Hillel, D. 1980. *Fundamentals of soil physics*. Academic Press, New York.

Jamison, V.C. 1953. Changes in air-water relationships due to structural improvement of soils. *Soil Sci.* 76:143–151.

Jamison, V.C. 1960. Water movement restriction by plant residues in a silt loam soil. *Trans. 7th Int. Congr. Soil Sci.* 1:464–472.

Jamison, V.C., and E.M. Kroth. 1958. Available moisture storage capacity in relation to textural composition and organic matter content of several Missouri soils. *Soil Sci. Soc. Am. Proc.* 22:189–192.

Jenny, H. 1930. The nitrogen content of the soil as related to the precipitation-evaporation ratio. *Soil Sci.* 29:193–206.

Joffe, J.S. 1955. Green manuring viewed by a pedologist. *Adv. Agron.* 7:141–187.

Kononova, M.M. 1966. *Soil organic matter: its nature, its role in soil formation and in soil fertility*. 2nd ed. Pergamon Press, New York.

Leuken, H., W.L. Hutcheon, and E.A. Paul. 1962. The influence of nitrogen on the decomposition of crop residues in the soil. *Can. J. Soil Sci.* 42:276–286.

Mann, H.H. 1959. Field studies in green manuring, II. *Emp. J. Exp. Agr.* 27:243–251.

Martin, J.P., and S.J. Richards. 1969. Influence of the copper, zinc, iron and aluminum salts of some microbial and plant polysaccharides on aggregation and hydraulic conductivity of Ramona sandy loan. *Soil Sci. Soc. Am. Proc.* 33:421–423.

Martin, J.P., W.P. Martin, J.B. Page, W.A. Raney, and J.D. De Ment. 1955. Soil aggregation. *Adv. Agron.* 7:1–37.

Millette, G.J.F., R.G. Leger, and R.G. Bailey. 1980. A comparative study of physical and chemical properties of two cultivated and forested Orthic Podzols. *Can. J. Soil Sci.* 60:707–719.

Morachan, Y.B., W.C. Moldenhauer, and W.E. Larson. 1972. Effects of increasing amounts of organic residues on continuous corn. I. Yields and soil physical properties. *Agron. J.* 64:199–203.

Mortensen, J.L., and J.L. Young. 1960. Effects of summer-seeded green manures on soil structure and availability of nitrogen. *Ohio Agr. Exp. Sta. Res. Bull.* No. 865.

Page, J.B., and C.J. Willard. 1947. Cropping systems and soil properties. *Soil Sci. Soc. Am. Proc.* 11:81–88.

Peele, T.C., O.W. Beale, and F.F. Lesnesne. 1948. Irrigation requirements of South Carolina soils. *Agr. Eng.* 29:157–158.

Pieters, A.J. 1927. *Green manuring*. John Wiley and Sons, New York.

Pieters, A.J., and R. McKee. 1938. Green manures. In: *Soils and men—the yearbook of agriculture, 1938*. USDA, Washington, DC. pp. 431–444.

Pinck, L.A., F.E. Allison, and V.L. Gaddy. 1946. The nitrogen requirement in the utilization of carbonaceous residues in soils. *J. Am. Soc. Agron.* 38:410–420.

Poyser, E.A., R.A. Hedlin, and A.O. Ridley. 1957. The effect of farm and green manures on the fertility of Blackearth-Meadow clay soils. *Can. J. Soil Sci.* 37:48–56.

Prince, A.L., S.J. Toth, A.W. Blair, and F.E. Bear. 1941. Forty-year studies of nitrogen fertilizers. *Soil Sci.* 52:247–261.

Pringle, J., and J.R.H. Coutts. 1956. The effect of grasses on aggregation in light soils. *J. Brit. Grass. Soc.* 11:185–189.

Ram, D.N., and P.J. Zwerman. 1960. Influence of management systems and cover crops on soil physical conditions. *Agron. J.* 52:473–476.

Reeve, M.J., P.D. Smith, and A.J. Thomasson. 1973. The effect of density on water retention properties of field soils. *J. Soil Sci.* 24:356–367.

Rennie, D.A., E. Truog, and O.N. Allen. 1954. Soil aggregation as influenced by microbial gums, level of fertility and kind of crop. *Soil Sci. Soc. Am. Proc.* 18:399–403.

Russell, E.W. 1973. *Soil conditions and plant growth.* 10th ed. Longman, London.

Sheard, R.W. 1977. Forages in a rotation. *Notes Agr.* 13(2):11–13.

Singh, A. 1962. Studies on the *modus operandi* of green manures in tropical climates. I. A critical review of literature. *Indian J. Agron.* 7:69–79.

Soane, B.D. 1975. Studies on some soil physical properties in relation to cultivations and traffic. In: Soil physical conditions and crop production. *Min. Agr. Fish and Food Tech. Bull.* No. 29, London. pp. 160–182.

Sopher, C.D., and R.J. McCracken. 1973. Relationships between soil properties, management practices and corn yields on South Atlantic Coastal Plain soils. *Agron. J.* 65:595–599.

Sowden, F.J. 1968. Effect of long-term annual additions of various organic amendments on the nitrogenous components of a clay and sand. *Can. J. Soil Sci.* 48:331–339.

Sowden, F.J., and A.J. Atkinson. 1968. Effect of long-term annual additions of various organic amendments on the organic matter of a clay and a sand. *Can. J. Soil Sci.* 48:323–330.

Stevenson, D.S. 1974. Influence of peat moss on soil water retention for plants. *Can. J. Soil Sci.* 54:109–110.

Strickland, E. 1951. The effect of soybeans on volume weight, and water stability of aggregates, soil organic matter content and crop yield. *Soil Sci. Soc. Am. Proc.* 15:30–34.

Tamboli, P.M., W.E. Larson, and M. Ameniya. 1964. Influence of aggregate size on soil moisture retention. *Iowa Acad. Sci.* 71:103–108.

Warman, P.R. 1980. The basics of green manuring. *Macdonald J.* 41(1):3–6.

Wischmeier, W.H., and J.V. Mannering. 1965. Effect of organic matter content of the soil on infiltration. *J. Soil Water Conserv.* 20:150–152.

Wisniewski, A.J., M. Salomon, and J.B. Smith. 1958. Restoration of soil structure with red top and other sod crops. *Soil. Sci. Soc. Am. Proc.* 22:320–322.

# The Influence of Macropores on the Transport of Dissolved and Suspended Matter Through Soil

R.E. White*



## I. Introduction

This article is concerned with the predominantly vertical movement of water through soils that to some degree have a network of large channels (macropores) and the consequences of this movement for the convective transport of solutes and suspended matter. Beven and Germann (1982) have reviewed the experimental evidence indicating that infiltration and redistribution of water in soils containing macropores are not adequately described by theories that treat the soil as a homogeneous medium conforming to Darcian principles of water flow. Such theories, developed

*Soil Science Laboratory, Department of Agricultural Science, University of Oxford, OX1 3PF, England

for the mixing of solutions in hollow tubes (Taylor, 1953) and porous rock strata (Brigham *et al.*, 1961), have been applied to miscible displacement experiments with columns of sand, resins, glass beads, or finely sieved and repacked soil. The underlying assumptions are that an unreactive solute moves through the medium at the same velocity as the water and all the resident water in the column is displaced by the incoming water. The analysis and interpretation of these experiments have been reviewed several times (Gardner, 1965; Biggar and Nielsen, 1967; Biggar and Nielsen, 1980; Wagenet, 1983) and will not be repeated here.

However, in undisturbed naturally structured soils, especially those with macropores, only partial displacement of resident water and solutes by incoming water and solutes occurs. Recognition of such behavior, variously described as "channeling," "short-circuiting," "bypassing," "preferential flow," or "partial displacement" (Scotter, 1978; Bouma, 1981; Beven and Germann, 1982; Smettem *et al.*, 1983) is not new (see Schumacher, 1864; Lawes *et al.*, 1882), as pointed out by Thomas *et al.* (1978). But only recently has attention been focused on this phenomenon because of the possible consequences for rapid leaching of fertilizers, soluble and particulate waste materials, and other potential pollutants into surface waters and ground water (Thomas and Phillips, 1979; Bouma *et al.*, 1983).

## II. Characterization of Macroporosity in Soil

Macropores are formed in various ways: by shrinkage at natural planes of weakness on drying (Brewer, 1964); chemical weathering leading to solution piping (Reeves, 1980); freeze–thaw cycles, mole draining, and subsoiling (Beven and Germann, 1982); plant roots (Aubertin, 1971); and soil fauna (Green and Askew, 1965; Ehlers, 1975). Beven and Germann (1982) concluded that most soils contained some macropores, the nature and volume of which depended on a dynamic balance between constructive and destructive processes. Thus, soil tillage (subsoiling excepted) generally destroys macropores (Thomas and Phillips, 1979) even though the bulk density of the disturbed soil is decreased and hence its total porosity increased (Cassel *et al.*, 1974; Kanchanasut *et al.*, 1978). The abundance of earthworm burrows is less in cultivated than in uncultivated soil (Ehlers, 1975; Abbott *et al.*, 1979), although Ehlers (1975) found that burrow numbers were comparable at 60 cm depth. Quisenberry and Phillips (1978) suggested that the existence of some macropores below plough depth (15 cm) in a cultivated Maury soil, a clayey, mixed, mesic Thermic Paleudult, was responsible for rapid but erratic flow of water into the subsoil.

The shape of macropores varies from planar slits (cracks or fissures), through voids of irregular cross section (vughs) to cylindrical pipes. Progress is now being made on quantifying the frequency distribution of pores of various shape using image-analyzing computers (Bullock and

Thomasson, 1979). Using thin sections of soil, the shape and area of voids in predetermined size classes above a minimum width of 1 μm are recorded. A three-dimensional picture of the pore space can be laboriously constructed from serial sections at 50-μm intervals, but ways of extrapolating mathematically from two to three dimensions are currently being explored (Ringrose-Voase and Bullock, 1984). Because there is normally no sharp break in the size distribution of pores in soil, the choice of a lower size limit for macropores (expressed as an equivalent cylindrical diameter) is arbitrary. Summarizing the literature, Beven and Germann (1982) reported minimum equivalent diameters for macropores ranging from 30 to 3000 μm.

Macroporosity has also been defined in terms of the pore volume that drains at a particular tension; for example, pores draining at a tension of 0.05 bars, corresponding to an equivalent diameter of 60 μm (Hall *et al.*, 1977). Alternatively, macropore volume has been identified from a sharp change in hydraulic conductivity in the low suction range as soil drains from a saturated state (Germann and Beven, 1981). For large undisturbed blocks of a soil derived from Oxford Clay (Evesham series), the latter authors reported macroporosities of 0.01 to 0.045 $cm^3$ $cm^{-3}$. Ehlers (1975) and Omoti and Wild (1979) made direct measurements of the number and size of earthworm burrows, while others have counted macropores in sections of resin-impregnated blocks of soil (Wilkins *et al.*, 1977; Hewitt and Dexter, 1980).

These estimates of macroporosity are not necessarily related to the efficacy of macropores in conducting water. Consequently, other approaches have been followed to identify the functional macroporosity of a soil. Topp and Davis (1981) studied the infiltration of water into soil cracks using time-domain reflectometry and found that the water content increased much faster close to the cracks than at some distance from them. In many cases dyes such as methylene blue, lissamine-yellow, or rhodamine B have been mixed with the water to indicate the water's pathway and the depth to which it has penetrated (Ritchie *et al.*, 1972; Anderson and Bouma, 1973; Omoti and Wild, 1979; Smettem and Trudgill, 1983). In other dye studies, the area of pore walls wet by infiltrating water has been measured (Bouma and Dekker, 1978), from which, if the pore widths are known, the volume of functional macropores can be calculated. Bouma and Wosten (1979) calculated by microscopic examination of two swelling clay soils that the macropore volume was 0.014 and 0.024 $cm^3$ $cm^{-3}$, of which about one-third was stained by methylene blue.

Although macropores may comprise only a small fraction of the total soil volume (0.001 to 0.05), they can have a profound effect on the rate of infiltration and redistribution of water under the right conditions. Beven and Germann (1982) suggest that rainfall or irrigation rates between 1 and 10 mm $hr^{-1}$ may be sufficient to initiate macropore flow, the exact value depending primarily on the saturated hydraulic conductivity of the peds comprising the soil matrix. Assuming laminar flow induced by gravity in vertical macropores that are filled with water, the rate of flow through individual channels is given by the equations (Childs, 1969)

$$Q_c = \left(\frac{\pi \rho g}{8\eta}\right) r^4 \qquad [1]$$

for cylinders and

$$Q_s = \left(\frac{\rho g}{12\eta}\right) d^3 \qquad [2]$$

for planar silts (cracks). $Q_c$ and $Q_s$ have the dimensions $L^3T^{-1}$, respectively, and $\rho$ is the fluid density, $g$ the acceleration due to gravity, $\eta$ the viscosity, $r$ the cylinder radius, and $d$ the crack width. Using equation [1], one can calculate the flow rate in individual pores within the macropore size range and the potential infiltration rate into a soil when macropores occupy either 0.01 or 0.05 of the surface area, assuming that the surface between the macropores makes no contribution to infiltration (Table 3.1). Clearly, as the macropore diameter increases above 20 μm, the infiltration rate potentially attributable to macropores rises rapidly. However, Ehlers (1975) measured an average flow rate of $3 \times 10^{-8}$ $m^3s^{-1}$ for individual earthworm burrows of 2000 to 5000 μm diameter which suggests that under natural conditions, rates of flow in macropores are about two orders of magnitude less than the hypothetical values in Table 1. This would apply particularly to the larger pores that are unlikely to fill completely with water, except when water is ponded on the surface. Then, however, air entrapment in the soil may restrict water entry (Dixon and Linden, 1972) while wall roughness and constrictions will reduce flow rates in macropores under all natural conditions.

**Table 1.** Hypothetical Flow Rates in Individual Cylindrical Macropores and Potential Infiltration Rates into Soil

| Pore diameter (μm) | Draining tension (bars) | Flow rate in one pore ($m^3\ s^{-1}$) | Potential infiltration ($mm\ hr^{-1}$) into a soil surface having a fractional macropore area of 0.01 | 0.05 |
|---|---|---|---|---|
| 2000 | 0.0015 | $3.8 \times 10^{-6}$ | $4.4 \times 10^4$ | $2.2 \times 10^5$ |
| 1000 | 0.003 | $2.4 \times 10^{-7}$ | $1.1 \times 10^4$ | $5.5 \times 10^4$ |
| 200 | 0.015 | $3.8 \times 10^{-10}$ | $4.4 \times 10^2$ | $2.2 \times 10^3$ |
| 20 | 0.15 | $3.8 \times 10^{-14}$ | $4.4 \times 10^0$ | $2.2 \times 10^1$ |

## III. Effects of Macropore Flow on the Transport of Applied Tracers

### A. Steady State

Tritiated water (THO) or non-adsorbed ions such as $Cl^-$ or $Br^-$ have been used as tracers for water flow in soil cores, lysimeters, and field soils (Cassel *et al.*, 1974; Bouma and Anderson, 1977; Addiscott *et al.*, 1978; Jury *et al.*, 1982). In some experiments, the depth to which the tracer penetrated was measured and shown to exceed the depth it should have reached if ideal miscible displacement had occurred (Blake *et al.*, 1973; Quisenberry and Phillips, 1976; Thomas *et al.*, 1978). In others, the ratio of tracer concentrations in the effluent from an undisturbed soil column ($C$) to tracer concentration in the input ($C_0$) was plotted against drainage, usually expressed as pore volumes of effluent ($V/V_0$). The resultant breakthrough curves can be examined for any effects of preferential water flow on tracer transport (Bouma, 1981). The analysis of the breakthrough curve for a step-change or pulse input of tracer under steady-state conditions using a convective-dispersive model (Biggar and Nielsen, 1980) assumes that the velocity of water in the pores is uniform over distances large compared to the characteristic pore size. The pore water velocity $v$ is calculated from

$$v = \frac{q}{\theta}, \qquad [3]$$

where $q$ is the water flux density (dimensions $LT^{-1}$) and $\theta$ the volumetric water content of the column. As a non-reactive tracer moves through the soil at velocity $v$, the spread or dispersion of its concentration about the average depth of displacement at any time $t$ is attributed to small velocity variations between pores and the diffusion of tracer between regions of different concentration. Terms for adsorption, precipitation, plant uptake, or biological transformation can be included in the analysis as required. The term $D$ in the basic equation,

$$\frac{\delta C}{\delta t} = D\frac{\delta^2 C}{\delta z^2} - v\frac{\delta C}{\delta z}, \qquad [4]$$

is a measure of the extent of mixing between the incoming and resident solutions and is called a dispersion coefficient (dimensions $L^2T^{-1}$). $z$ is the distance in the vertical direction. A typical breakthrough curve following a step-change in input of an unreactive tracer, for which the underlying assumptions of the convective-dispersive equation would apply, is shown as curve I in Figure 1.

However, Biggar and Nielsen (1962) noted that even in relatively homogeneous porous media (columns of packed soil aggregates) an unquantified "diffusion-velocity dispersion interaction" occurred. In such

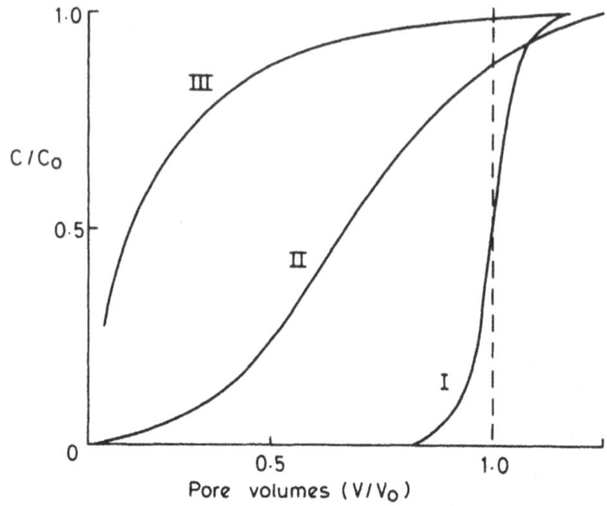


**Figure 1.** Types of breakthrough curve for a step-change in input of a non-reactive solute; I - ideal miscible displacement; II, III - partial displacement of resident soil water owing to anion-exclusion, dead-end pores, or a wide spread of pore water velocities.

cases, *D* is essentially a curve-fitting parameter (Wild and Cameron, 1980) and can vary through several orders of magnitude, depending on the mean velocity, its distribution, and the size of the structural units (Passioura, 1971; Passioura and Rose, 1971). Moreover, experiments with undisturbed soils usually show tracer breakthrough occurring much earlier, with a differently shaped curve from that of curve I (Kissel *et al.*, 1973; Bouma and Anderson, 1977; Bouma *et al.*, 1983; White *et al.*, 1984). Similarly, pulse-input experiments on undisturbed soils sometimes (Tyler and Thomas, 1981) but not always (Cassel *et al.*, 1974) show a peak concentration in the effluent much earlier than one pore volume. Curves II and III in Figure 1 are examples of breakthrough curves following a step-change in input of an unreactive tracer to undisturbed soil cores. Type II curves can be described by modifying equation [4] to allow for the effects of anion-exclusion and dead-end pores (Elprince and Day, 1977) or by subdividing the soil water into "mobile" and "immobile" or "stagnant" regions (van Genuchten and Wierenga, 1976; Gaudet *et al.*, 1977). It is doubtful, however, whether the convective-dispersive equation, even when modified in this way, is appropriate for the description of solute transport in soils in which there is a wide distribution of pore water velocities (Rose, 1973; Davidson *et al.*, 1983). This is the case for soils with macropores.

Bresler and Dagan (1979) concluded that the spread of solute concentration in a field soil resulting from lateral variations in hydraulic conductivity under constant flux leaching was very much greater than the spread caused by pore-scale hydrodynamic dispersion. Where the spatial variability of soil structure renders the various mechanisms affecting solute transport almost unknowable, Jury (1982) has argued, as an alternative to mechanistic modeling, the use of a transfer function based on the probability density function of solute travel times through the soil (Raats, 1978). In this case, all of the dispersion of the applied solute is attributed to velocity variations in the soil. When the mean and standard deviation of the velocity distribution have been determined for a particular depth, the distribution of solute at greater depths and longer times can be accurately predicted provided there is no marked change in the physical properties of the soil with depth (Jury *et al.*, 1982). The evidence available suggests that the distribution of pore water velocities during constant flux irrigation of field soils is approximately log-normal (Biggar and Nielsen, 1976; van der Pol *et al.*, 1977). White *et al.* (1984) assumed a log-normal distribution of pore water velocity to analyze the breakthrough of step inputs of $Cl^-$ and THO under constant flux irrigation onto large (20 cm diameter × 25–30 cm long) undisturbed cores from a range of Kentucky soils. Examples of the experimental and model breakthrough curves for $Cl^-$, which were of type II or III (Figure 1), are shown in Figure 2. Mean pore water velocities $\bar{v}$ and the standard deviations of ln $v$ were obtained from the curves of best fit. The fractional volume of water participating in solute transport ($\theta_m$) was estimated from

$$\theta_m = \frac{q}{\bar{v}} \quad [5]$$

and ranged from 25% of the soil water in a well-structured Crider silt loam to 93% in a recently tilled Bruno sandy loam (Ohio River flood plain). This suggested that the incoming water was flowing down macropores and bypassing much of the resident soil water so that the tracers were diffusing from the "mobile" water into the "immobile" water of the matrix. The faster diffusing tracer (THO) was retarded relative to $Cl^-$, as first observed by Biggar and Nielsen (1962), and its breakthrough curve displaced further to the right of the $Cl^-$ curve than would be predicted from anion-exclusion effects alone (McMahon and Thomas, 1974).

Further evidence of preferential flow down macropores was provided by the comparison of $Cl^-$ breakthrough with that of antibiotic-resistant *Escherichia coli*, applied in suspension to the top of undisturbed cores (Smith *et al.*, 1985). As shown in Figure 3, $C/C_0$ for the bacteria, which were effectively excluded from pores <15 μm diameter during the 10 to 12 hr of these experiments (White, 1985b), remained constant during drainage. The average $C/C_0$ for the whole experiment reflected the extent to which the organisms could be transported through the larger pores. On the other hand, $C/C_0$ for $Cl^-$ steadily approached one as Cl-enriched water, moving at different velocities through a wide range of pore sizes, contributed to the column efffluent. White (1985b) showed that 69% of the variation in mean

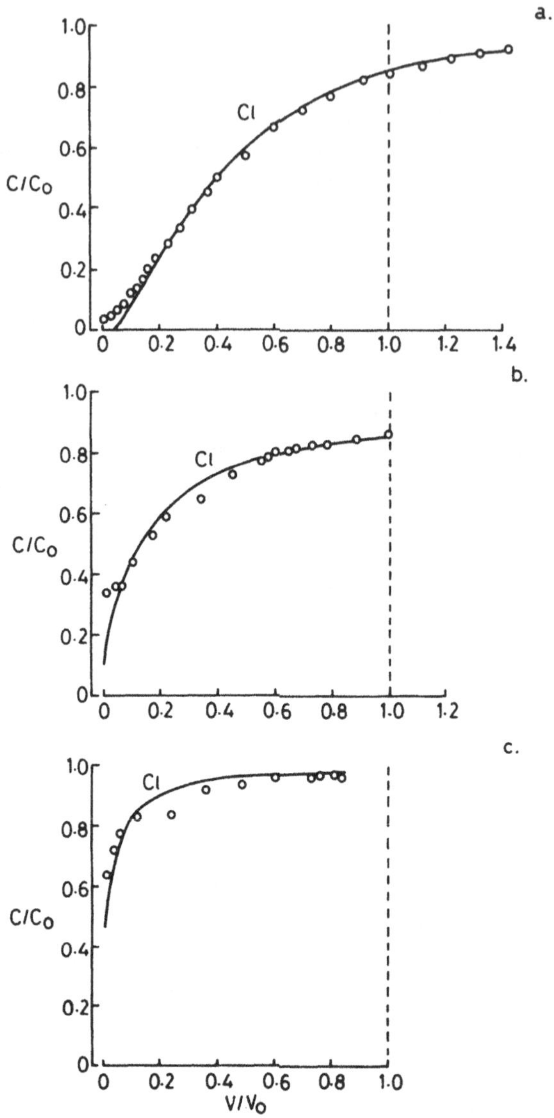


**Figure 2.** Breakthrough curves for a step-change in $Cl^-$ input to three cores of undisturbed soil (a) Maury silt loam, (b) Crider silt loam, (c) Bruno sandy loam. Input rate 18 to 20 mm $hr^{-1}$. Fitted lines derived from a transfer function model (after White *et al.*, 1984).

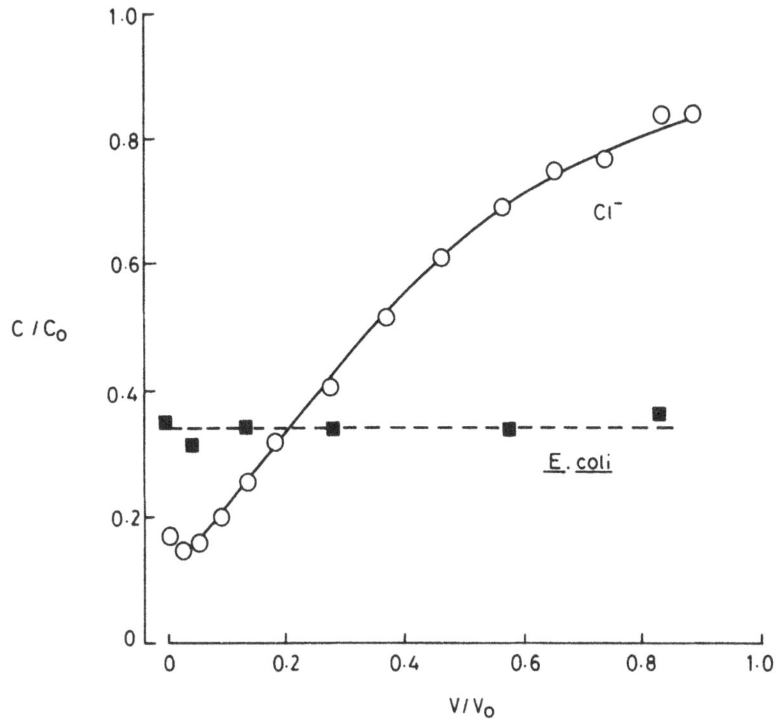


**Figure 3.** Breakthrough curves for $Cl^-$ and antibiotic-resistant *Escherichia coli* in an undisturbed Maury silt loam. Input rate 20 mm $hr^{-1}$ (after Smith *et al.*, 1985). Reproduced by permission of the American Society of Agronomy, Crop Science Society of America, Soil Science Society of America.

$C/C_0$ for *E. coli* could be accounted for by regression on mean pore water velocity in the undisturbed soils. But Smith *et al.* (1985) found that when the continuity of macropores was destroyed by disturbing part or all of the column, $C/C_0$ for *E. coli* was much reduced (Table 2). Rahe *et al.* (1978) attributed the movement of *E. coli* at a speed of 15 m $hr^{-1}$ through a near saturated hillslope soil to preferential flow through macropores.

There are fewer reports of the effect of macropores on the breakthrough of reactive solutes such as the cations and phosphate in soil. Shaffer *et al.* (1979) recorded immediate large increases in the concentration of $Cd^{2+}$ in drainage sampled from macropores at 1.2 m depth in a soil to which $CdCl_2$ solutions were applied. $Cd^{2+}$ concentrations in the soil matrix water, sampled through suction cups, with one exception remained low throughout the experiment. Using small cores (5.4 cm diameter × 10 cm long) of a soil comprising 96% sand, Mansell *et al.* (1977) were able to predict reasonably well the breakthrough of a pulse input of $H_2PO_4^-$ that peaked in the effluent after 2 to 10 pore volumes. Their convective-dispersive model included terms for reversible adsorption, irreversible precipitation, and a

**Table 2.** Convective Transport of *E. coli* Through Disturbed and Undisturbed Soil Cores[a]

| Soil | Mean $C/C_0$ during breakthrough | |
|---|---|---|
| | Undisturbed | Disturbed |
| Crider silt loam, a fine-silty, mixed, mesic Typic Paleudalf | 0.44 | 0.07 |
| Maury silt loam, a clayey, mixed, mesic Typic Paleudult | 0.22 | 0.002 |
| Bruno sandy loam, a sandy, mixed, thermic Typic Udifluvent (Kentucky River site) | 0.79 | 0.05 |

[a]Cores taken from grassland sites; irrigation rate 20 mm $hr^{-1}$ (after Smith *et al.*, 1985). Reproduced by permission of the American Society of Agronomy, Crop Science Society of America, Soil Science Society of America.

dispersion coefficient derived from $Cl^-$ breakthrough. But Kanchanasut *et al.* (1978) obtained $H_2PO_4^-$ breakthrough after only 0.05 pore volumes in a small core of a well-structured soil. The rapid breakthrough of $H_2PO_4^-$, attributed to the presence of a worm hole, could be simulated using puddled soil pierced by a single cylindrical channel or planar slit. Kanchanasut *et al.* (1978) also reported a ten-fold increase in P concentration in tile drain flow from a soil when rainfall followed immediately the application of superphosphate. This is consistent with results obtained for a drained Evesham clay soil near Oxford (Hargreaves, 1983). Scotter (1978) developed a model to describe the breakthrough of both sorbed and non-sorbed solutes in saturated cores containing large holes or cracks of known dimensions. Laminar flow of solution under a gravitational potential gradient in the channels was coupled with diffusion (with or without reaction) from the channels into the non-conducting matrix.

### B. Unsteady State

The effects of macropore flow on solute transport in initially dry soils when unsteady, unsaturated flow occurs are less well defined. Smiles and coworkers (Smiles *et al.*, 1978; Smiles and Gardiner, 1982) have developed the theory of infiltration and solute dispersion in dry soils, but only for columns of packed, finely sieved soils. The extent of bypass flow in undisturbed soils depends on their structure, the hydraulic conductivity of the peds, the application rate, and the soil water content (Bouma, 1981). When the application rate exceeds the saturated hydraulic conductivity of the peds in the matrix, water flows under the influence of gravity through the interped channels even though the water potential may be too low for them to fill. Dye stains in bands 2.5 to 7.5 mm wide in the subsoil of cracking clay soils (Bouma and Dekker, 1978) suggested film flow down channel walls (Beven and Germann, 1981). This and other studies (Bouma and Wosten, 1979; Hoogmoed and Bouma, 1980) indicated that the actual contact area between the flowing water and ped faces in dry cracking clays was extremely variable.

For very dry soils (matric potentials of $-15$ bars or so) there will be no outflow from the matrix for some time, and the ratio of outflow to inflow rate can be termed the "bypassing ratio" (Kneale and White, 1984). Bouma *et al.* (1981) reported ratios of 0.36 to 0.47 for dry clay soils under grass at input rates of 17 to 25 mm $hr^{-1}$. Kneale and White (1984) found the bypassing ratio increased with input rate for a dry cracked Evesham soil once a threshold of 2.2 mm $hr^{-1}$, the saturated hydraulic conductivity of the matrix, was exceeded (Table 3). The effect of antecedent soil water content on $Cl^-$ transport is illustrated in Figure 4 for undisturbed cores of Maury silt loam at input rates of 20 mm $hr^{-1}$ (White *et al.*, 1984). The drier the soil initially, the more marked the bypass flow as indicated by higher $C/C_0$ values in the early part of the breakthrough curves. Clearly, when no resident soil water is displaced the effluent concentration will be reduced only by diffusion of $Cl^-$ into the immobile water of the matrix. Breakthrough curves of this type have been consistently observed in this laboratory using initially dry cores and application rates of 7 to 14 mm $hr^{-1}$ (non-ponding conditions). The analysis of these curves to estimate the change with time in the volume of soil water participating in solute transport is being investigated (White, 1985a), since it is crucial to the modeling of nitrate leaching from soil during intermittent rainfall events (see below).

**Table 3.** Bypassing Ratio as a Function of Application Rate for Dry Cores of Grassland Soil (Evesham Series)[a]

| Application rate (mm $hr^{-1}$) | Matrix absorption rate at steady state outflow (mm $hr^{-1}$) | Bypassing ratio |
|---|---|---|
| 2.2 | 2.2 | 0 |
| 3.6 | 2.7 | 0.25 |
| 11.9 | 6.6 | 0.44 |
| 21.9 | 9.5 | 0.55 |

[a]Soil initially at matric potential $\simeq -15$ bars; $\theta$ in the range 0.26–0.30 $cm^3\ cm^{-3}$ (after Kneale and White, 1984).

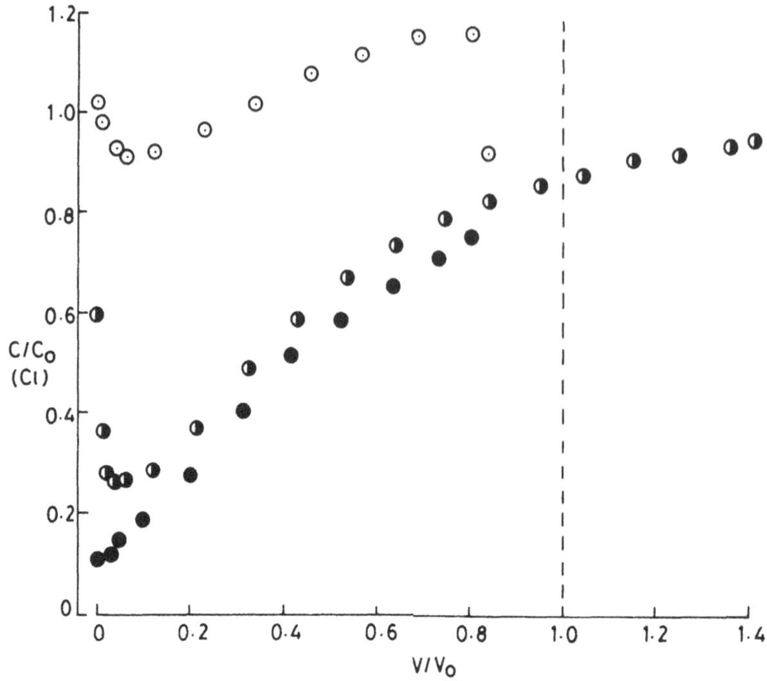


**Figure 4.** Breakthrough curves for a step-change in Cl input to three cores of undisturbed Maury silt loam of different initial water contents. Input rate 20 mm $hr^{-1}$. ○ $\theta = 0.18$; ◑ $\theta = 0.34$; ● $\theta = 0.38$ (after White *et al.*, 1984).

## IV. Effect of Macropore Flow on the Transport of Indigenous Solutes in Soil

Much of the data on this aspect derives from studies on the leaching of soil nitrate and the displacement of salts from saline soils. These are discussed in turn.

### A. Nitrate Leaching

The effect of macropores on nitrate leaching is complex not only because of uncertainty about the relative contribution of different pathways to water flow but also because of the variation in nitrate concentration over short distances. Within a soil block of 1 $m^3$, the uneven distribution of N fertilizer, excreta from animals, and of mineral N derived from soil organic matter creates spatial heterogeneity. On a smaller scale, the dynamic balance betwen nitrification and denitrification can vary with oxygen status across the width of a soil aggregate (Tiedje *et al.*, 1984).

Cunningham and Cooke (1958), Wild (1972), and Thomas and Phillips (1979) have suggested that when nitrate is held within soil aggregates it will be protected from leaching when bypass flow occurs. Others (Addiscott and Cox, 1976; Barraclough *et al.*, 1983) have observed that if nitrate fertilizer has recently been applied to the soil, or if soil-generated nitrate is held on the outside of aggregates, then bypassing flow causes it to leach faster than it would by uniform miscible displacement. Certainly, water draining from soils that are rewetting after prolonged dry weather often has a high concentration of nitrate (Williams, 1976; Garwood and Tyson, 1977; Foster and Walling, 1978). Martin and White (1982) and Macduff and White (1984) reported high nitrate concentrations coinciding with peak discharges from a partially underdrained clay catchment, especially during autumn or after N fertilizer was applied in spring. White *et al.* (1983) attributed the high autumn concentrations to a combination of (1) a flush of mineralization of soil organic N, especially in the surface 20 cm, as the dry soil rewetted (Birch, 1958), and (2) rapid flow of nitrate-enriched water during storms from the surface to mole and tile drains, bypassing the unsaturated soil matrix of low nitrate concentration at depths of 40 to 60 cm. Leeds-Harrison *et al.* (1982) have suggested a similar flow mechanism in mole-drained soil, and Barraclough *et al.* (1983) have obtained comparable results for nitrate leaching in an underdrained soil. Moling and subsoiling increase the number and continuity of macropores (Godwin *et al.*, 1981), and there is evidence that nitrate concentrations in tile flow are higher than in seepage flow (Thomas and Barfield, 1974). This is contrary to the general belief that nitrate concentrations will be lower, the faster water flows through the soil (Letey *et al.*, 1977).

Shaffer *et al.* (1979) compared the nitrate concentration of water sampled by suction cups (matrix flow) at several depths in an undisturbed soil with that of water draining from macropores at 1.2 m. Water with and without nitrate was applied intermittently to the surface. The macropore drainage responded immediately to a change in concentration of the applied water (increase or decrease), whereas the concentration in the matrix flow showed only a small gradual increase with time. It was estimated that 90% of the total flow in this soil occurred in pores $\geq$ 150 μm diameter. This was one of the more comprehensive studies of the effect of a dual pore system on nitrate leaching. Further evidence for the effect of macropore flow on nitrate leaching is provided by Shuford *et al.* (1977). They found that nitrate was removed from all depths down to 1 m in a well-drained silt loam following single irrigations with 89 and 178 mm of water (Figure 5). The recovery of nitrate down to 1.35 m was also less than predicted by a convective-dispersive model, suggesting that some nitrate had moved below this depth.

Nevertheless, as discussed in the next section, it is also held that leaching of solutes from soil is more efficient under the unsaturated conditions prevailing during intermittent flow than under the saturated conditions induced by ponding (Biggar and Nielsen, 1967; Shainberg and Oster, 1978). The relevance of this view to the leaching of soil nitrate can be

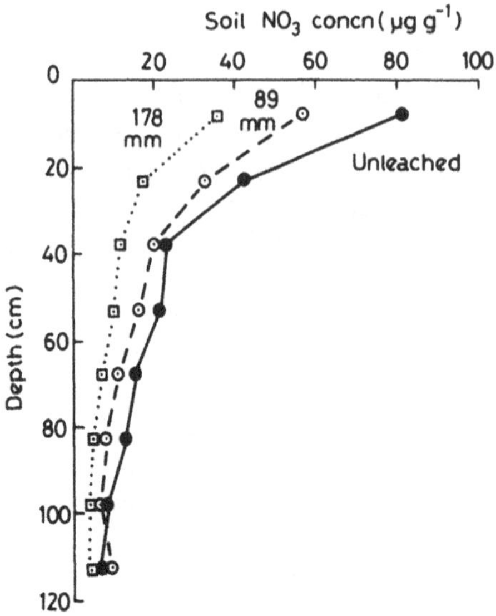


**Figure 5.** Redistribution of soil nitrate following single applications of 89 and 178 mm of irrigation water to a well-drained silt loam (after Shuford *et al.*, 1977).

evaluated better when more is known about the diffusive flux of nitrate across the interface between water within aggregates and bypassing water in the macropores. Addiscott (1977) assumed no diffusive interchange during flow but complete equilibration between mobile and immobile water between events. Subsequently, Addiscott (1984) incorporated solute diffusion between the mobile and immobile regions during flow into his model and concluded that the presence of solute in an immobile phase retards leaching because diffusion between phases never reaches equilibrium. Darrah *et al.* (1983) calculated that for an intra-aggregate nitrate concentration of 14 µg N $cm^{-3}$, nitrate could diffuse to a sink of zero concentration in the macropores at the rate of 2.3 µg N $cm^{-2}$ of contact area per hour. Using a minimum value of vertical contact area of 80 $cm^2$ per cm depth per $m^2$ (Bouma and Dekker, 1978), this amounts to 0.04 kg N $ha^{-1}$ $h^{-1}$ for the top 20 cm of soil, suggesting that removal of nitrate from the region of immobile water may not be insignificant. Bouma and coworkers quote contact areas on ped faces from 80 to 1635 $cm^2$ per cm depth per horizontal area of 1 $m^2$. White *et al.* (1984), using a double tracer technique, estimated the perimeter of conducting channels in undisturbed cores to be 1350 to 1650 cm per $m^2$ of cross section (equivalent to contact area per cm depth of soil). There is a need for more data on contact areas between mobile and immobile water during bypass flow in other undisturbed soils.

### B. Salt Leaching

Few papers refer explicitly to the role of macropores in salt leaching. This may reflect the fact that many such studies have been done on irrigated soils of light texture and weakly developed structure, to which convective-dispersive models have been successfully applied (Biggar and Nielsen, 1967).

Miller *et al.* (1965) compared the efficiency with which surface-applied KCl was leached by continuous or intermittent ponding in a Panoche clay loam. The average velocity of flow under intermittent ponding, when matrix flow should have predominated, was only one-twentieth that under continuous ponding, when macropore flow should have occurred. However, 50% more $Cl^-$ was removed by the former treatment from the top 90 cm of soil after 610 mm of water had been applied in both cases. On the other hand, Cameron and Wild (1982a) found no significant difference between the profile distributions of $^{36}Cl^-$ in a silty clay loam over chalk after 80 mm of irrigation (two inputs of 40 mm) or 71 mm of intermittent net rainfall. Further study of this question is required, with particular emphasis being placed on the extent to which macropore flow occurs under the two treatments.

Nevertheless, in poorly drained, naturally saline soils, the creation of greater macroporosity can have a pronounced effect on the rate of salt leaching (McIntyre *et al.*, 1982a, 1982b). In their experiment, one block (6.5 × 4.5 × 4.5 m deep) of gypsum-ameliorated sodic clay soil was ponded with water for 145 days and an adjacent block of untreated soil for 379 days. The water content of the untreated soil did not change below 2.4 m, whereas 265 mm out of the 605 mm of water that infiltrated the treated plot penetrated below 4.5 m. Virtually all the original soil $Cl^-$ was leached from above 2 m in the gypsum-treated soil, compared to a net leaching of only 50% above 1 m in the untreated soil and net accumulation of $Cl^-$ between 1 and 2.3 m (see Figure 6). The difference between the two plots was attributed to greater convective transport through macropores and diffusion of $Cl^-$ out of the matrix in the gypsum-treated soil. The increased $Ca^{2+}$ concentration provided by the gypsum would have reduced the anion-exclusion effect which Blackmore (1976) found restricted the rate of salt diffusion out of peds in a similar saline, cracking clay soil.

## V. Discussion

A general conclusion to be drawn from studies on undisturbed soils is that macropores can greatly decrease the time taken for dissolved and suspended matter applied to the surface to reach subsurface drains or ground water (Thomas and Phillips, 1979; Hagedorn *et al.*, 1981; Bouma *et al.*, 1983). However, the effect of macropores on the *quantity* of solute

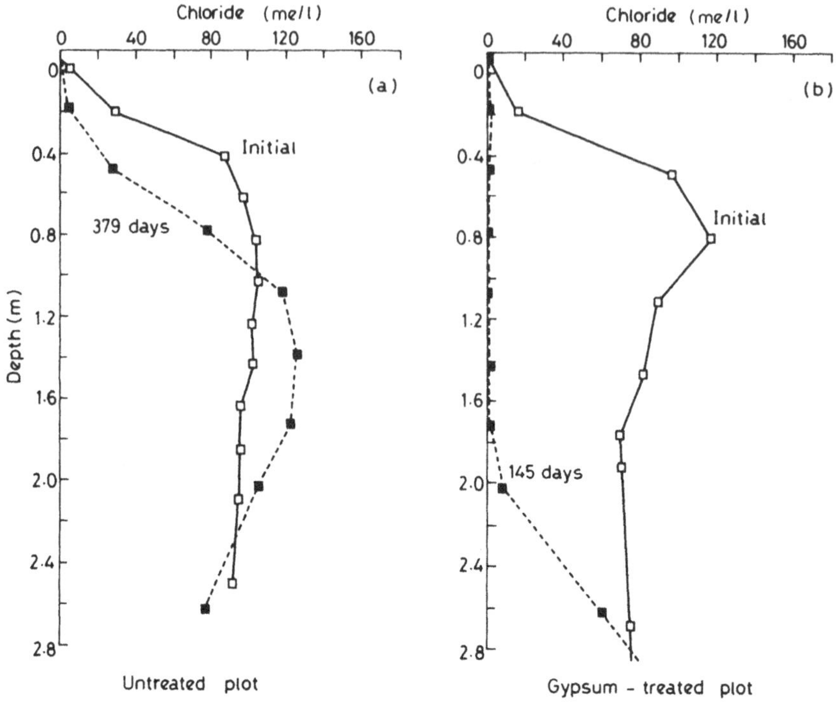


**Figure 6.** Redistribution of soil chloride in a sodic cracking clay with and without macropores. (a) untreated plot, ponded for 379 days; (b) gypsum-treated plot, ponded for 145 days (after McIntyre *et al.*, 1982b).

leached from soil per millimeter of water applied is less easily prescribed. Whether macropore flow that bypasses a fraction of the resident water within soil peds increases or decreases the amount of solute leached depends on:

1. the location of the solute (whether recently applied fertilizer or a soil-generated solute held on the exterior or within peds);
2. the ratio of macropore to matrix flow;
3. the saturated hydraulic conductivity of the matrix;
4. the antecedent water content of the soil;
5. the contact area between the bypass flow and the relatively static water of the matrix; and
6. the rate of solute diffusion between the mobile and immobile water volumes.

Further research is required on all these aspects.

A good test of the level of understanding of the processes involved in leaching is how well models can simulate experimental results. Solutions of the convective-dispersive equation (equation [4]) are sometimes successful in predicting solute redistribution with depth in field soils (Gardner, 1965;

Cameron and Wild, 1982b), or the leaching of solute from undisturbed cores and lysimeters (Cassel *et al.*, 1975; Mansell *et al.*, 1977), although deviations of predicted from observed values are often attributed to macropore flow (Rose *et al.*, 1982). In applying equation [4] it is usually assumed that the soil water content remains constant so that the macroscopic pore water velocity ($v$) is fixed and all of the variability in solute transport is incorporated in $D$, the dispersion coefficient. This can be resolved further according to the equation (Cameron and Wild, 1982b; Jensen, 1983).

$$D = f D_0 + \lambda v, \qquad [6]$$

where $D_0$ is the diffusion coefficient of the solute in bulk solution, $f$ an impedance factor, and $\lambda$ is a variable called the *dispersivity*, defined by equation [6]. For a column of packed impermeable granules of uniform size, $\lambda$ should correlate with particle size (Jensen, 1983), but in heterogeneous media such as undisturbed soils, $\lambda$ is variable and must be obtained by curve-fitting. Values of $\lambda$ ranging from 0.1 to 20 cm have been obtained. If this approach is to be pursued, some way of making an independent estimate of $\lambda$ needs to be found, possibly based on an index of soil structure. However, the problem becomes unwieldy for unsteady unsaturated flow, which is common enough in field soils, because $v$, $\lambda$, and $f$ in equation [6] then vary.

Another approach to modeling leaching in field soils involves subdividing the soil profile into layers with all (Burns, 1974) or part (Addiscott, 1977; Addiscott *et al.*, 1978) of the soil water participating in transport. The rate of water flow down the profile is governed by the net rainfall rate and the storage capacity of each layer. However, when tested on soils other than those for which they were developed, these models have not been very successful (Cameron and Wild, 1982b; Barraclough *et al.*, 1983). Addiscott's model belongs to the family of models that divide the soil water into "mobile" and "immobile" regions (van Genuchten and Wierenga, 1976; Rao *et al.*, 1980). These suffer from the disadvantage that the mobile volume must be arbitrarily defined or the transfer coefficient between the two regions obtained by curve-fitting. From this basic concept, Skopp *et al.* (1981) developed a model specifically for soils with macropores in which two mobile regions (corresponding to interped and intraped pathways) are recognized, each characterized by a porosity, flow velocity, and dispersion coefficient. Solute flux between the two regions is calculated from the concentration difference and an interaction term that depends on ped geometry and thickness of the intraped region. Although theoretically the model can predict breakthrough curves of type III (Figure 1), it too has the disadvantage that several of the parameters cannot be independently measured.

The pronounced spatial variability of soil hydraulic properties often poses severe problems for the application of convective-dispersive and layer models to field soils. Where a large number of measurements of, for example, hydraulic conductivity and soil water content have been made at various sites in the field, an average curve relating conductivity to water content can be generated by using scaling theory and lumping all the soil variability into a single scaling factor (Nielsen *et al.*, 1982). The distribution of this scaling factor must then be determined. However, a promising alternative to these models, the value of which is limited by how well the mechanisms of water flow and solute transport in field soils are understood, is provided by the transfer function model (Jury, 1982, 1983). This requires only that the mass of solute transported be conserved and that the probability density function of solute travel times between the input and output surfaces in the soil be known. If the density function of travel times, or alternatively pore water velocities, conforms to a simple function such as a log-normal distribution, then the mean and standard deviation of that distribution are the essential parameters of the model. A log-normal distribution was applicable for steady flow through cores of undisturbed soil ~20 cm in diameter (White *et al.*, 1984), as well as for field plots 0.01 ha in area (Biggar and Nielsen, 1976). Even for 930 mm of intermittent rainfall and irrigation on a loamy sand in California, a log-normal distribution seemed to hold, enabling Jury *et al.* (1982) to predict the transport of $Br^-$ to 1.8 m depth following a pulse input of NaBr to the surface. To extend this model further, more data are required on the distributions of pore water velocity in soils, especially those with macropores and under conditions when flow is not steady. The application of this approach to the leaching of indigenous soil solutes such as nitrate also needs to be explored.

## VI. Summary

This article has briefly reviewed the nature and extent of large continuous channels (macropores) in undisturbed soils and examined the evidence for macropores acting as conduits for the rapid infiltration and redistribution of water containing dissolved and suspended matter. The breakthrough of externally applied solutes and the leaching of indigenous solutes such as nitrate and salts were discussed for conditions of steady and unsteady flow. The convective-dispersive theory of solute transport has limitations in predicting the redistribution of solutes and their appearance in the drainage from soils with macropores. Other approaches to modeling solute breakthrough in undisturbed soils, including the use of a transfer function based on the probability density function of solute travel times, were discussed. The effects of intermittent vs. continuous leaching and of the distribution of solutes between macropores and micropores in the soil matrix were examined in respect of the efficiency with which salt and nitrate can be removed from soils. It is concluded that more quantitative information on the conditions controlling macropore flow and on the probability density function of solute travel times in undisturbed soils is needed.

## Acknowledgments

I am grateful to Mrs. J. Tree for preparation of the diagrams and to Dr W.R. Kneale for the data in Table 1.

## List of Symbols (Other Than Chemical Formulae and Abbreviations)

| | |
|---|---|
| $C$, $C_0$ | Solute concentration in effluent and in input water respectively ($M\,L^{-3}$) |
| $d$ | Crack width ($L$) |
| $D_0$ | Diffusion coefficient in bulk solution ($L^2T^{-1}$) |
| $D$ | Dispersion coefficient ($L^2T^{-1}$) |
| $f$ | Impedance factor (dimensionless) |
| $g$ | Acceleration due to gravity ($LT^{-2}$) |
| $\eta$ | Viscosity ($ML^{-1}\,T^{-1}$) |
| $\theta$ | Volumetric water content ($L^3\,L^{-3}$) |
| $\theta_m$ | Fractional soil volume participating in solute transport ($L^3\,L^{-3}$) |
| $\lambda$ | Dispersivity ($L$) |
| $\pi$ | The constant "Pi" (dimensionless) |
| $q$ | Water flux density ($LT^{-1}$) |
| $Q_c$, $Q_s$ | Water flow rate in a cylinder and planar silt, respectively ($L^3\,T^{-1}$) |
| $\rho$ | Fluid density ($M\,L^{-3}$) |
| $r$ | Cylindrical pore radius ($L$) |
| $v$, $\bar{v}$ | Pore water velocity and mean pore water velocity respectively ($LT^{-1}$) |
| $t$ | time ($T$) |
| $V$ | Cumulative effluent volume ($L^3$) |
| $V_0$ | Soil pore volume ($L^3$) |
| $z$ | Vertical distance, positively downwards ($L$) |

## References

Abbott, I., C.A. Parker, and I.D. Sills. 1979. Changes in the abundance of large soil animals and physical properties of soils following cultivation. *Austr. J. Soil Res.* 17:343–353.

Addiscott, T.M. 1977. A simple computer model for leaching in structured soils. *J. Soil Sci.* 28:554–563.

Addiscott, T.M. 1984. Modelling the interaction between solute leaching and intraped diffusion in clay soils. In: J. Bouma and P.A.C. Raats (eds.), *Water and solute movement in heavy clay soils*, ILRI Publ. No. 37, Wageningen pp. 279–292.

Addiscott, T.M., and D. Cox. 1976. Winter leaching of nitrate from autumn-applied calcium nitrate, ammonium sulphate, urea and sulphur-coated urea in bare soil. *J. Agr. Sci., Camb.* 87:381–389.

Addiscott, T.M., D.A. Rose, and J. Bolton. 1978. Chloride leaching in the Rothamsted drain gauges: influence of rainfall pattern and soil structure. *J. Soil Sci.* 29:305–314.

Anderson, J.L., and J. Bouma. 1973. Relationships between saturated hydraulic conductivity and morphometric data of an argillic horizon. *Soil Sci. Soc. Am. Proc.* 37:408–413.

Aubertin, G.M. 1971. The nature and extent of macropores in forest soils and their influence on subsurface water movement. *USDA For. Serv. Res. Paper* NE-192.

Barraclough, D., M.J. Hyden, and G.P. Davies. 1983. Fate of fertilizer nitrogen applied to grassland. I. Field leaching results. *J. Soil Sci.* 34:483–497.

Beven, K., and P. Germann. 1981. Water flow in soil macropores. II. A combined flow model. *J. Soil Sci.* 32:15–30.

Beven, K., and P. Germann. 1982. Macropores and water flow in soils. *Water Resour. Res.* 18:1311–1325.

Biggar, J.W., and D.R. Nielsen. 1962. Miscible displacement. II. Behaviour of tracers. *Soil Sci. Soc. Am. Proc.* 26:125–128.

Biggar, J.W., and D.R. Nielsen. 1967. Miscible displacement and leaching phenomena. In: R.M. Hagan, H.R. Haise, and T.W. Edminster (eds.), *Irrigation of agricultural lands*. Agron. Monog. No. 11. American Society of Agronomy, Madison. pp. 254–274.

Biggar, J.W., and D.R. Nielsen. 1976. Spatial variability of the leaching characteristics of a field soil. *Water Resour. Res.* 12:78–84.

Biggar, J.W., and D.R. Nielsen. 1980. Mechanisms of chemical movement in soils. In: A. Banin and U. Kafkafi (eds.), *Agrochemicals in soils*. Pergamon, Oxford. pp. 213–227.

Birch, H.F. 1958. The effect of soil drying on humus decomposition and nitrogen availability. *Plant Soil* 10:9–31.

Blackmore, A.V. 1976. Salt sieving within clay soil aggregates. *Austr. J. Soil Res.* 14:149–158.

Blake, G., E. Schlichting, and U. Zimmerman. 1973. Water recharge in a soil with shrinkage cracks. *Soil Sci. Soc. Am. J.* 37:669–672.

Bouma, J. 1981. Soil morphology and preferential flow along macropores. *Agr. Water Manage.* 3:235–250.

Bouma, J., and J.L. Anderson. 1977. Water and chloride movement through soil columns simulating pedal soils. *Soil Sci. Soc. Am. J.* 41:766–770.

Bouma, J., and L.W. Dekker. 1978. A case study on infiltration into dry clay soil. I. Morphological observations. *Geoderma* 20:27–40.

Bouma, J., and J.H.M. Wosten. 1979. Flow patterns during extended saturated flow in two undisturbed swelling clay soils with different macrostructures. *Soil Sci. Soc. Am. J.* 43:16–22.

Bouma, J., J.W. Dekker, and C.J. Muilwijk. 1981. A field method for measuring short-circuiting in clay soils. *J. Hydrol.* 52:347–354.

Bouma, J., C. Belmans, L.W. Dekker, and W.J.M. Jeurissen. 1983. Assessing the suitability of soils with macropores for subsurface liquid waste disposal. *J. Environ. Qual.* 12:305–310.

Bresler, E., and G. Dagan. 1979. Solute dispersion in unsaturated heterogeneous soil at field scale. II. Applications. *Soil Sci. Soc. Am. J.* 43:467–472.

Brewer, R. 1964. *Fabric and mineral analysis of soils*. John Wiley and Sons, New York.

Brigham, W.E., P.W. Reed, and J.N. Dew. 1961. Experiments on mixing during miscible displacement in porous media. *Soc. of Petrol. Eng. J.* 1:1–8.

Bullock, P., and A.J. Thomasson. 1979. Rothamsted studies of soil structure. II. Measurement and characterization of macroporosity by image analysis and comparison with data from water retention measurements. *J. Soil Sci.* 30:391–414.

Burns, I.G. 1974. A model for predicting the redistribution of salts applied to fallow soils after excess rainfall or evaporation. *J. Soil Sci.* 25:165–178.

Cameron, K.C., and A. Wild. 1982a. Comparative rates of leaching of chloride, nitrate and tritiated water under field conditions. *J. Soil Sci.* 33:649–657.

Cameron, K.C., and A. Wild. 1982b. Prediction of solute leaching under field conditions: an appraisal of three methods. *J. Soil Sci.* 33:659–669.

Cassel, D.K., M.T. van Genuchten, and P.J. Wierenga. 1975. Predicting anion movement in disturbed and undisturbed soils. *Soil Sci. Soc. Am. Proc.* 39:1015–1019.

Cassel, D.K., T.H. Kruegar, F.W. Schroer, and E.B. Norum. 1974. Solute movement through disturbed and undisturbed soil cores. *Soil Sci. Soc. Am. Proc.* 38:36–40.

Childs, E.C. 1969. *An introduction to the physical basis of soil water phenomena*. John Wiley and Sons, New York.

Cunningham, R.K., and G.W. Cooke. 1958. Soil nitrogen. II. Changes in levels of inorganic nitrogen in a clay loam soil caused by fertilizer application, by leaching and uptake by grass. *J. Sci. Food Agr.* 9:317–324.

Darrah, P.R., P.H. Nye, and R.E. White. 1983. Diffusion of $NH_4^+$ and $NO_3^-$ mineralized from organic N in soil. *J. Soil Sci.* 34:693–707.

Davidson, J.M., P.S.C. Rao and P. Nkedi-kizza. 1983. Physical processes influencing water and solute transport in soils. In: D.W. Nelson, D.E. Elrick and

K.K. Tanji (eds.) Chemical mobility and reactivity in soil systems. SSSA Spec. Publ. No. 11, pp 35–47.

Dixon, R.M., and D.R. Linden. 1972. Soil air pressure and water infiltration under border irrigation. *Soil Sci. Soc. Am. Proc.* 36:948–953.

Ehlers, W. 1975. Observations on earthworm channels and infiltration on tilled and untilled loess soils. *Soil Sci.* 119:242–249.

Elprince, A.M., and P.R. Day. 1977. Fitting solute breakthrough equations to data using two adjustable parameters. *Soil Sci. Soc. Am. J.* 41:39–41.

Foster, I.D.L., and D.E. Walling. 1978. The effects of the 1976 drought and autumn rainfall on stream solute levels. *Earth Surf. Process.* 3:393–406.

Gardner, W.R. 1965. Movement of nitrogen in soil. In: W.V. Bartholomew and F.E. Clark (eds.), *Soil nitrogen* Am. Soc. Agron. Monogr. No.10, American Society of Agronomy, Madison. pp. 550–572.

Garwood, E.A., and K.C. Tyson. 1977. High loss of nitrogen in drainage from soil under grass following a prolonged period of low rainfall. *J. Agr. Sci., Camb.* 89:767–768.

Gaudet, J.P., H. Jegat, G. Vachaud, and P.J. Wierenga. 1977. Solute transfer, with exchange between mobile and stagnant water, through unsaturated sand. *Soil Sci. Soc. Am. J.* 41:665–671.

Germann, P., and K. Beven. 1981. Water flow in soil macropores. I. An experimental approach. *J. Soil Sci.* 32:1–13.

Godwin, R.J., G. Spoor, and P. Leeds-Harrison. 1981. An experimental investigation into the force mechanics and resulting soil disturbance of mole ploughs. *J. Agr. Eng. Res.* 26:477–497.

Green, R.D., and G.P. Askew. 1965. Observations on the biological development of macropores in soils of Romney Marsh. *J. Soil Sci.* 16:342–349.

Hagedorn, C., E.L. McCoy, and T.M. Rahe. 1981. The potential for ground-water contamination from septic effluents. *J. Environ. Qual.* 10:1–8.

Hall, D.G.M., M.J. Reeve, A.J. Thomasson, and V.F. Wright. 1977. Water retention, porosity and density of field soils. *S.S.E.W. Tech. Monogr.* No.9, Harpenden.

Hargreaves, L. 1983. The sources and sinks of inorganic phosphorus in a stream flowing from a calcareous soil. D.Phil. thesis, University of Oxford.

Hewitt, J.S., and A.R. Dexter. 1980. Effects of tillage and stubble management on the structure of a swelling soil. *J. Soil Sci.* 31:203–215.

Hoogmoed, W.B., and J. Bouma. 1980. A simulation model for predicting infiltration into cracked clay soil. *Soil Sci. Soc. Am. J.* 44:458–461.

Jensen, J.R. 1983. Chloride dispersion in packed columns during saturated steady flow. *J. Soil Sci.* 34:249–262.

Jury, W.A. 1982. Simulation of solute transport using a transfer function model. *Water Resour. Res.* 18:363–368.

Jury, W.A. 1983. Chemical transport modelling: current approaches and unresolved problems in: D.W. Nelson, D.E. Elrick and K.K. Tanji (eds.) Chemical mobility and reactivity in soil systems. SSSA Spec. Publ. No. 11, pp. 49–64.

Jury, W.A., L.H. Stolzy, and P. Shouse. 1982. A field test of the transfer function model for predicting solute transport. *Water Resour. Res.* 18:369–375.

Kanchanasut, P., D.R. Scotter, and R.W. Tillman. 1978. Preferential solute movement through larger soil voids. II. Experiments with saturated soil. *Austr. J. Soil Res.* 16:269–276.

Kissel, D.E., J.T. Ritchie, and E. Burnett. 1973. Chloride movement in undisturbed clay soil. *Soil Sci. Soc. Am. Proc.* 37:21–24.

Kneale, W.R., and R.E. White. 1984. The movement of water through cores of dry (cracked) clay loam grassland topsoil. *J. Hydrol.* 67:361–365.

Lawes, J.B., J.H. Gilbert, and R. Warington. 1882. *On the amount and composition of the rain and drainage waters collected at Rothamsted.* William Clowes and Sons, London.

Leeds-Harrison, P., G. Spoor, and R.J. Godwin. 1982. Water flow to mole drains. *J. Agr. Eng. Res.* 27:81–91.

Letey, J., J.W. Blair, D. Devitt, L.J. Lund, and P. Nash. 1977. Nitrate-nitrogen in effluent from agricultural tile drains in California. *Hilgardia* 45:289–319.

Macduff, J.H., and R.E. White. 1984. Components of the nitrogen cycle measured for cropped and grassland soil-plant systems. *Plant Soil* 76:35–47.

McIntyre, D.S., J. Loveday, and C.L. Watson. 1982a. Field studies of water and salt movement in an irrigated swelling clay soil. I. Infiltration during ponding. *Austr. J. Soil Res.* 20:81–90.

McIntyre, D.S., J. Loveday, and C.L. Watson. 1982b. Field studies of water and salt movement in an irrigated swelling clay soil. III. Salt movement during ponding. *Austr. J. Soil Res.* 20:101–105.

McMahon, M.A., and G.W. Thomas. 1974. Chloride and tritiated water flow in disturbed and undisturbed soil cores. *Soil Sci. Soc. Am. Proc.* 38:727–732.

Mansell, R.S., H.M. Selim, P. Kanchanasut, J.M. Davidson, and J.G.A. Fiskell. 1977. Experimental and simulated transport of phosphorus through sandy soils. *Water Resour. Res.* 13:189–194.

Martin, R.P., and R.E. White. 1982. Automatic sampling of stream water during storm events in small remote catchments. *Earth Surf. Proc. and Land.* 7:53–61.

Miller, R.J., D.R. Nielson, and J.W. Biggar. 1965. Chloride displacement in Panoche clay loam in relation to water movement and distribution. *Water Resour. Res.* 1:63–73.

Nielsen, D.R., J.W. Biggar, and P.J. Wierenga. 1982. Nitrogen transport processes in soil. In: F.J. Stevenson (ed.), *Nitrogen in agricultural soils*. Am. Soc. Agron. Monogr. No. 22. American Society of Agronomy, Madison. pp. 423–448.

Omoti, U., and A. Wild. 1979. Use of fluorescent dyes to mark the pathways of solute movement through soils under leaching conditions. 2. Field experiments. *Soil Sci.* 128:93–104.

Passioura, J.B. 1971. Hydrodynamic dispersion in aggregated media. I. Theory. *Soil Sci.* 111:339–344.

Passioura, J.B., and D.A. Rose. 1971. Hydrodynamic dispersion in aggregated media. 2. Effects of velocity and aggregate size. *Soil Sci.* 111:345–351.

Quisenberry, V.L., and R.E. Phillips. 1976. Percolation of surface-applied water in the field. *Soil Sci. Soc. Am. J.* 40:484–489.

Quisenberry, V.L., and R.E. Phillips. 1978. Displacement of soil water by simulated rainfall. *Soil Sci. Soc. Am. J.* 42:675–679.

Raats, P.A.C. 1978. Convective transport of solutes by steady flows. I. General theory. *Agr. Water Manage.* 1:201–218.

Rahe, T.H., C. Hagedorn, E.L. McCoy, and G.F. Kling. 1978. Transport of antibiotic resistant *Escherichia coli* through a western Oregon hillslope soil under conditions of saturated flow. *J. Environ. Qual.* 7:487–494.

Rao, P.S.C., D.E. Rolston, R.E. Jessup, and J.M. Davidson. 1980. Solute transport in aggregated porous media: theoretical and experimental evaluation. *Soil Sci. Soc. Am. J.* 44:1139–1146.

Reeves, M.J. 1980. Recharge of the English Chalk. A possible mechanism. *Eng. Geol.* 14:231–240.

Ringrose-Voase, A.J. and Bullock, P. 1984. The automatic recognition and measurement of soil pore types by image analysis and computer programs. *J. Soil Sci.* 35:673–684.

Ritchie, J.T., D.E. Kissel, and E. Burnett. 1972. Water movement in undisturbed swelling clay soil. *Soil Sci. Soc. Am. Proc.* 36:874–879.

Rose, C.W., F.W. Chichester, J.R. Williams, and J.T. Ritchie. 1982. Application of an approximate analytic method of computing solute profiles with dispersion in soils. *J. Environ. Qual.* 11:151–155.

Rose, D.A. 1973. Some aspects of the hydrodynamic dispersion of solutes in porous materials. *J. Soil Sci.* 24:284–295.

Schumacher, W. 1864. *Die physik des Bodens*. Berlin.

Scotter, D.R. 1978. Preferential solute movement through larger soil voids. I. Some computations using simple theory. *Austr. J. Soil Res.* 16:257–267.

Shaffer, K.A., D.D. Fritton, and D.E. Baker. 1979. Drainage water sampling in a wet, dual-pore soil system. *J. Environ. Qual.* 8:241–246.

Shainberg, I., and J.D. Oster. 1978. Quality of irrigation water. *Inter. Irrig. Information Centre, Publ.* No.2, Bet-Dagan.

Shuford, J.W., D.R. Fritton, and D.E. Baker. 1977. Nitrate-nitrogen and chloride movement through undisturbed field soil. *J. Environ. Qual.* 6:255–259.

Skopp, J., W.R. Gardner, and E.J. Tyler. 1981. Solute movement in structured soils: two region model with small interaction. *Soil Sci. Soc. Am. J.* 45:837–842.

Smettem, K.R.J., and S.T. Trudgill. 1983. An evaluation of some fluorescent and non-fluorescent dyes in the identification of water transmission routes in soils. *J. Soil Sci.* 34:45–56.

Smettem, K.R.J., S.T. Trudgill, and A.M. Pickles. 1983. Nitrate loss in soil drainage waters in relation to by-passing flow and discharge on an arable site. *J. Soil Sci.* 34:499–509.

Smiles, D.E., and B.N. Gardiner. 1982. Hydrodynamic dispersion during unsteady unsaturated water flow in a clay soil. *Soil Sci. Soc. Am. J.* 46:9–14.

Smiles, D.E., J.R. Philip, J.H. Knight, and D.E. Elrick. 1978. Hydrodynamic dispersion during absorption of water by soil. *Soil Sci. Soc. Am. J.* 43:229–234.

Smith, M.S., G.W. Thomas, R.E. White, and D. Ritonga. 1985. Transport of *Escherichia coli* through intact and disturbed columns of soil. *J. Environ. Qual.* 14: 87–91.

Taylor, G.L. 1953. Dispersion of soluble matter in solvent flowing slowly through a tube. *Proc. Roy. Soc.* 219:186–203.

Tiedje, J.M., A.J. Sexstone, T.B. Parkin, N.P. Revsbech, and D.R. Shelton. 1984. Anaerobic processes in soil. *Plant Soil* 76:197–212.

Thomas, G.W., and B.J. Barfield. 1974. The unreliability of tile effluent for monitoring subsurface nitrate-nitrogen losses from soils. *J. Environ. Qual.* 3:183–185.

Thomas, G.W., and R.E. Phillips. 1979. Consequences of water movement in macropores. *J. Environ. Qual.* 8:149–152.

Thomas, G.W., R.E. Phillips, and V.L. Quisenberry. 1978. Characterization of water displacement in soils using simple chromatographic theory. *J. Soil Sci.* 29:32–37.

Topp, G.C., and J.L. Davis. 1981. Detecting infiltration of water through soil cracks by time-domain reflectometry. *Geoderma* 26:13–23.

Tyler, D.D., and G.W. Thomas. 1981. Chloride movement in undisturbed soil columns. *Soil Sci. Soc. Am. J.* 45:459–461.

Van der Pol, R.M.P., P.J. Wierenga, and D.R. Nielsen. 1977. Solute movement in a field soil. *Soil Sci. Soc. Am. J.* 41:10–13.

van Genuchten, M.T., and P.J. Wierenga. 1976. Mass transfer in sorbing porous media. I. Analytical solutions. *Soil Sci. Soc. Am. Proc.* 40:473–480.

Wagenet, R.J. 1983. Principles of salt movement in soils. In: D.W. Nelson, D.E. Elrick and K.K. Tanji (Eds.) Chemical Mobility and Reactivity in Soil Systems. SSSA Spec. Publ. No. 11 pp 123–140.

White, R.E. 1985a. The analysis of solute breakthrough curves to predict water redistribution during unsteady flow through undisturbed structured clay soil. *J. Hydrol.* (in press).

White, R.E. 1985b. Transport of chloride and non-diffusible solutes in soils. *Irrig. Sci.* 6:3–10.

White, R.E., G.W. Thomas, and M.S. Smith. 1984. Modelling water flow through undisturbed soil cores using a transfer function model derived from $^{3}HOH$ and Cl transport. *J. Soil Sci.* 35:159–168.

White, R.E., S.R. Wellings, and J.P. Bell. 1983. Seasonal variations in nitrate leaching in structured clay soils under mixed land use. *Agr. Water Manage.* 7:391–410.

Wild, A. 1972. Nitrate leaching under bare fallow at a site in northern Nigeria. *J. Soil Sci.* 23:315–324.

Wild, A., and I.A. Babiker. 1976. The asymmetric leaching patterns of nitrate and chloride in a loamy sand under field conditions. *J. Soil Sci.* 27:460–466.

Wild, A., and K.C. Cameron. 1980. Soil nitrogen and nitrate leaching. In: P.B. Tinker (ed.), Soils and agriculture. *Critical Reports on Applied Chemistry. Vol. 2.* Blackwell Scientific Publications, Oxford. pp. 35–70.

Wilkins, D.E., W.F. Buchele, and W.G. Lovely. 1977. A technique to index soil pores and aggregates larger than 20 µm. *Soil Sci. Soc. Am. J.* 41:139–140.

Williams, R.J.B. 1976. The chemical composition of rain, land drainage and bore hole water from Rothamsted, Broom's Barn, Saxmundham and Woburn Experimental Stations. In: Agriculture and water quality. *M.A.F.F. Tech. Bull.* No.32. H.M.S.O., London. pp. 174–200.

# Behavior of Herbicides in Irrigated Soils

B. Yaron,* Z. Gerstl,* and W. F. Spencer**



*Institute of Soils and Water, The Volcani Center, Bet-Dagan 50-250, Israel
**USDA, Agricultural Research Service, Soil and Environmental Science Department, UCR, Riverside, California 92521


Advances in Soil Science, Volume 3

## I. Introduction

The primary function of herbicides is to protect agricultural crops from infestation with weeds and to prevent arable land from being overgrown by plant cover indigenous to the ecosystem. The chemicals known as herbicides are mainly synthetic organic compounds with broad molecular configurations having as a common property the ability of selectively killing or inhibiting the growth of plants. A selective herbicide retards growth or kills one plant species (the weed), whereas another plant species (the crop) is unaffected by the same treatment.

Herbicides are applied directly to the plants or incorporated into the soil. When applied to plants, the herbicides move by symplastic translocation from cell to cell via the interconnecting protoplasmotic fluid until they enter the phloem. When applied to the soil the herbicides are absorbed by the roots and move in the plant by apoplastic translocation through the roots following the same pathway as water (Klingman and Ashton, 1982). Translocation from soil upwards follows the transpiration stream, and in this situation the soil moisture content becomes a governing factor. Translocation of most herbicides is not restricted either to symplast or apoplast, but may involve both despite the fact that many herbicides appear to be limited primarily to either one of the other systems. Once added to a system, a herbicide may be translocated into the plant, volatilized into the atmosphere, leached down below the rootzone to the ground water, adsorbed on soil constituents causing carry over of biocidal effect to the next crop, transported while adsorbed on solid particles, or degraded.

The properties of the soil–air–water system, as well as the properties of the herbicide itself, define the fate of the chemical molecule, its dissipation into and below the root zone, and its translocation into the plant. By changing the properties of the system—even those affecting only the soil moisture content—we are able to affect the herbicidal activity of the organochemical, its dissipation in the environment, and its distribution in the soil atmosphere or the unsaturated zone between the soil surface and the ground-water table. Irrigation practices therefore become tools in governing the soil capacity to affect the persistence of herbicides, their efficiency in crop protection, and their potential pollution of the groundwater.

Many review articles and books have summarized the existing knowledge in thc field of soil–herbicides interactions (e.g., Kearney and Kaufman, 1969; Goring and Hamaker, 1972; Guenzi, 1974; Hance, 1980a; Hartley and Graham-Bryce, 1980), but very few were devoted to the particular aspect of irrigated agriculture. The last International Congress of Pesticide Chemistry (Kyoto, Japan, 1982) included a symposium on herbicides for rice (Oryza sativa) culture. Comprehensive review articles like those of

Crosby (1983) and Kuwatsuka (1983) describe the fate of herbicides in paddy soils, but the consideration is limited to flooded conditions.

The aim of our paper is to discuss the factors and the processes affecting the fate of herbicides under various irrigation regimes and to provide a conceptual interpretation for the existing information on soil–water–herbicide interactions in general, and for the results obtained in experiments carried out under irrigation in particular. It is not the intention of the authors to exhaust all sources of knowledge by carrying out a comprehensive bibliographic survey. This review will concentrate on the principles of various processes and reactions rather than attempt to enumerate all the existing information from the literatuare.

## II. Factors to Be Considered

Five main factors may affect the behavior of herbicides in irrigated agriculture: (1) the chemical characteristics of the herbicides, (2) the properties of the soil medium and their spatial variability, (3) the herbicide application techniques, (4) the irrigation technology, and finally (5) the fluctuations in the environmental conditions during the irrigation season.

The chemical properties of a herbicide that are important are its solubility in water, its vapor pressure, and its chemical and biochemical stability to surface or biologically induced reactions. The composition of the solid and liquid phase of the soil medium defines the adsorption–desorption of the herbicide from and into the liquid phase. Transport of the chemical is affected by the soil environment. The chemical mobility and reactivity of the herbicide is a function of the natural variation of the soil mineralogical and chemical composition, as well as microscopic, macroscopic, and megascopic domains affecting the velocities of water and solutes. The fate of herbicides in the soil medium may be affected by both the application technique and the irrigation practice. Surface broadcasting may favor herbicide volatilization or transport by runoff, while application in irrigation water may result in distribution within the soil profile. The irrigation practice affects the distribution patterns of herbicides in soil which may or may not lead to an increase or decrease in the herbicidal efficiency accompanied by a leaching process and potential pollution of the ground-water. Irrigation also causes fluctuations in the water content and temperature of the soil medium which in turn directly affect herbicide behavior.

### A. Herbicide Properties

Knowledge of both the chemical structures and physico-chemical properties of herbicides is a determining factor in understanding their behavior in the soil–water system. In the framework of our paper, it is impossible to describe the large number of herbicides used. Since the present paper is addressed to soil scientists we felt it desirable to make a compromise—i.e., to extend the part dealing with herbicides' properties without going into details, and to illustrate the broad spectrum of properties characterizing the herbicides interacting with the soil.

The commonly used herbicides are grouped—based on their chemical structure—in a number of well-defined classes and their properties (such as molecular weight, solubility in water, and vapor pressure) are summarized in Appendix 1. The grouping of the herbicides follows that of Ashton and Crafts (1973) and Klingman and Ashton (1982).

A brief description of the major herbicides follows.

## 1. Aliphatics

This group includes glyphosate, clorinated aliphatic acids (TCA and dalapon), the organic arsenicals (cacodylic acid (MAA, MSMA, DSMN, and MAMA), methyl bromide, and acrolein. Glyphosate is active through the leaves but has little effect when applied to the soil. TCA and dalapon are usually formulated as sodium salts. TCA (trichloracetic acid) is translocated through the plant, mainly from the roots via the apoplastic system. Dalapon has a chemical structure similar to that of TCA, but one chlorine atom has been replaced by a methyl ($-CH_3$) group. Dalapon is a systemic herbicide involving both apoplastic and symplastic translocation. Cacodylic acid and its sodium salts are translocated in the apoplast and methanearsonic acid (MAA) and its formulated compounds are translocated in both the symplast and apoplast. A herbicide specific to aquatic weeds and injected into the water is acrolein (acrylaldehyde or 2-propenal).

## 2. Amides

The amid type herbicides are characterized by the structural formula:

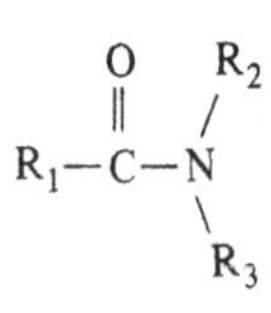


By substituting the radical groups, various amide herbicides are obtained. For example, chloroacetamides have a monochlorinated methyl group ($Cl-CH_2$) in the $R_1$ position of the amide structure. They are all applied directly to the soil. Some herbicides in this group are alachlor, butachlor, and propachlor. Some of the other amides used as herbicides have at least one ring structure in the $R_1$ position of the molecule. They are mainly soil-applied herbicides, and some of the compounds are diphenamid, pronamide, benzamide, and napropamide.

3. Benzoics

The benzoic herbicides are derivatives of benzoic acid

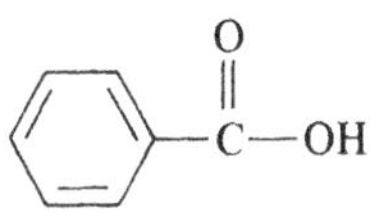


benzoic acid

containing chlorine atoms, methoxy, or amino groups. In general the benzoic herbicides are applied both to plants or soil. Dicamba and 2,3,6-TBA are two benzoic herbicides with similar chemical structures. 2,3,6-TBA is the common name for 2,3,6-trichlorobenzoic acid. In dicamba the chlorine atom at the number 2 position is replaced by a methoxy ($-OCH_4$) group. Both herbicides are characterized by apoplastic and symplastic transport.

4. Bipyridyliums

Those herbicides that belong to the bipyridylium quaternary ammonium class are characterized by a positive charge and are heterocyclic compounds.

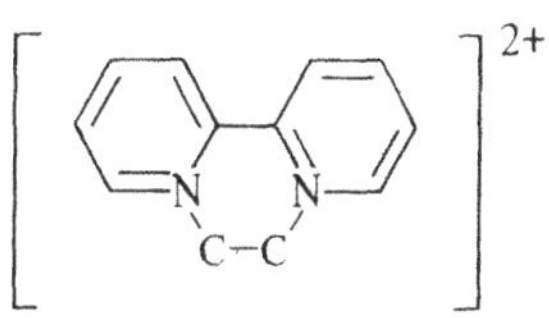


diquat ion

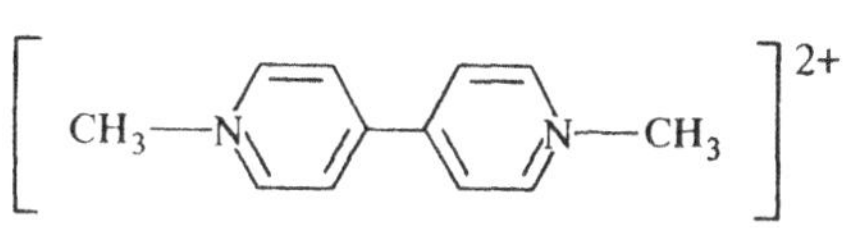


paraquat ion

Diquat and paraquat, the two herbicides of this group, are mainly contact herbicides but may be used also as aquatic biocides. The only translocation is via the apoplastic system.

5. Dinitroanilines

Chemical structure of this group is characteristic of anilines and 2,6-dinitroanilines.

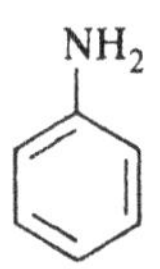


aniline

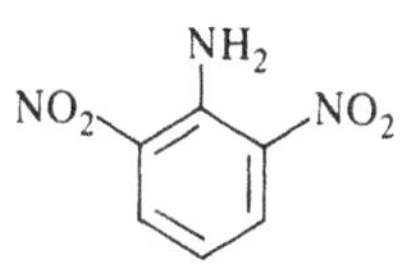


dinitroaniline

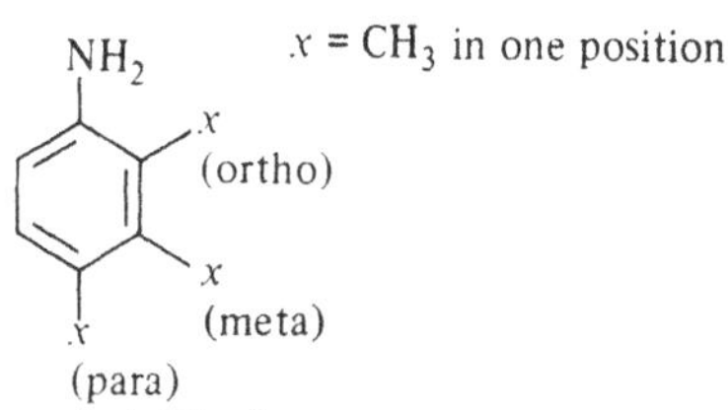


(toluidine)

A large number of dinitroaniline herbicides on the market are based on various radicals linked to the ring. In general, the herbicides of this group are quite volatile and are applied mainly as soil-incorporated herbicides. They are adsorbed by plants coming into contact with the treated soil, but translocation is limited.

Trifluralin is one of the most widely used herbicides from the dinitroaniline group. This herbicide inhibits growth of roots but is not significantly translocated into higher plants.

Other examples of herbicides in this group are: benefin, butralin, isopropalin, and pendimethalin.

### 6. Diphenyl Ethers

The diphenyl ether herbicides are used both as foliar and soil herbicides. They are mainly contact herbicides since little translocation occurs. Two diphenyl ethers will be used as examples: Acifluorfen is a contact herbicide but light is required to develop injury symptoms of the plant; nitrofen is essentially insoluble in water and is applied both as a foliar and a soil herbicide, but translocation into the plant is quite limited. Nitrofen is largely applied in cultivated paddy rice.

### 7. Benzonitriles

The herbicides included in this group are characterized by a $-C{\equiv}N$ group with $OH^-$, $Cl^-$, and/or $Br^-$ substitution on the benzene ring. These are applied both as soil and foliar herbicides.

Dichlobenil is generally incorporated into the soil to reduce its volatility and is absorbed by roots and translocated into plants via apoplast. Bromaxynil is a foliar herbicide that also has herbicidal effect when applied to the soil. Its translocation into the plant is limited and acts by direct injury to the plant tissue.

### 8. Phenoxys

The herbicides included in this group are characterized by the structural formula of:

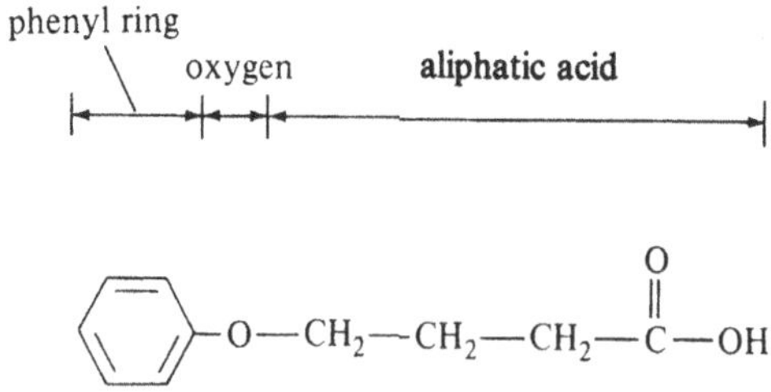


The phenyl ring is attached to an oxygen, attached to an aliphatic acid. The length of the carbon chain of the aliphatic acid defines the name of the herbicide. The phenoxys are formulated as emulsifiable acid, amino salts, mineral salts, or esters; their solubility in water and volatilization properties are governed by the type of formulation. 2,4-D is the best known herbicide of the phenoxy group. 2,4-D is used in foliar sprays and as a soil-applied pre-emergence herbicide. Roots adsorb 2,4-D in polar form and leaves in non-polar forms (acid and ester forms). Translocation is via both the symplastic and apoplastic systems. Translocation up or down is favored by a soil moisture content favorable to rapid plant growth. Low soil moisture contents may slow the herbicide translocation into the plant.

MCPA is similar to 2,4-D but has a methyl ($-CH_3$) group at the number 2 position of the ring replacing the chlorine atom. 2,4-DB and 2,4-DP are two other phenoxyl herbicides with similar biocidal properties to 2,4-D.

## 9. Carbamates and Thiocarbamates

Carbamate and thiocarbamate herbicides derive their base structural formula from carbamic acid:

Carbamate

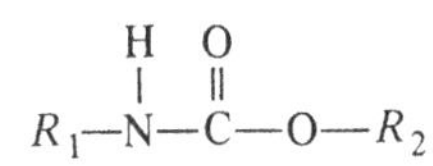

Thiocarbamate

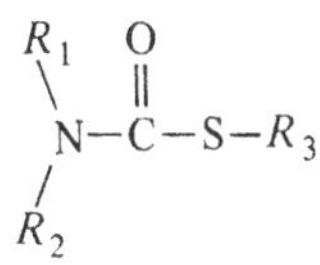

In the thiocarbamates, one of the oxygen atoms is replaced by a sulfur atom. Other substitutions also occur. The carbamates are soil applied (propham and chlorpropham) or foliar applied (e.g., asulam, phenmedipham). They are characterized by a high specific selectivity on particular crops. Propham applied to the soil is readily translocated to the plants when adsorbed by roots. Chlorpropham is more persistent in the soil than propham itself. Other carbamates like barban and phenmedipham are mainly applied to the foliage, and their interaction with the soil is of limited significance.

Thiocarbamates are usually applied to the soil. Their high volatility requires quick incorporation into the soil or application in irrigation water. Thiocarbamate pesticides are typified by butylate, which is a soil-incorporated compound translocated upward to shoots following root uptake via the apoplast system. EPTC is a volatile soil-incorporated herbicide that is adsorbed by seeds, roots, and emerging shoots in contact with treated soil. Metham (vapam) is formulated as a water-soluble solution. It is a temporary soil fumigant able to control germinating weed seeds and perennial weeds and is applied and incorporated into the soil.

## 10. Triazines

The triazines are heterocyclic nitrogen derivatives, the ring structure being composed of nitrogen and carbon atoms. Most triazines are symmetrical.

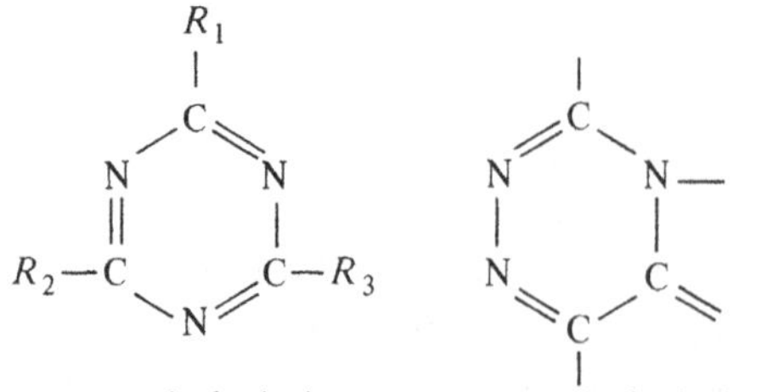


symmetrical triazine    asymmetrical triazine

The substituent at the $R_1$ position determines the ending of the common name. With chlorine atoms, the common name ends in *-azine*; with methylthio group, *-tryn*; and with methoxy group ($-OHC_3$), *-ton*. The solubility of the compound is defined by the $R_1$ substituent with the $-OCH_3$ substitution resulting in the highest solubility.

The triazine herbicides are adsorbed by leaves but are not translocated from them. When applied to the soil they are rapidly adsorbed by roots and translocated through the plants by the transpiration stream, almost exclusively in the apoplast system

The following compounds are arbitrarily selected as examples of triazines: Atrazine is widely used to control annual grasses but is effective also when applied to soil as a pre-emergence herbicide; prometryn is a selective herbicide used as pre-emergence and preplant soil incorporated; prometon is applied to the soil via incorporation or in irrigation water.

### 11. Substituted Ureas

Substituting three of the hydrogen atoms of urea with other chemical groups such as phenyl, methyl, and/or methoxy groups produces effective herbicides. The chemical structure of substituted urea is:

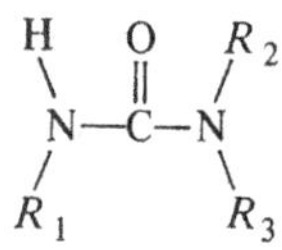


Most urea herbicides are nonselective and usually are applied to soil. They are adsorbed by roots and translocated apoplastically to the upper plant. The amount of herbicide translocated differs for various substituted urea compounds. Diuron is widely used as a pre-emergence herbicide for control of a large number of weeds. Monuron has a solubility in water much greater than diuron and was the first urea-type herbicide produced.

### 12. Uracils

The uracils are soil-applied herbicides having a bromine or chlorine atom in position 5 and a tert-butyl group in position 3. The uracil-type herbicides' mode of action is similar to that of the substituted ureas, and they are adsorbed by roots and translocated apoplastically to the plant tops. Bromacil and terbacil are two widely used uracil herbicides.

### 13. Other Herbicides

Other organic and a few inorganic herbicides are used in weed control. However, it is not our intent to cover the whole list of existing organic herbicides, Sodium chlorate, borates, and ammonium sulfamate are inorganic herbicides still used in weed control practices.

## B. Soil Characteristics: Properties and Heterogeneity

Both the solid and the liquid phases of the soil affect the fate of herbicides reaching the soil.

### 1. Properties

The two major soil components of significance to herbicide behavior and persistence are clay and organic matter contents. The liquid phase composition can affect conversions of the applied chemicals and their movement and distribution in the soil. It is not our aim to discuss the properties of the soil clays and organic matter. Comprehensive monographs on these subjects have been published (e.g., Grim, 1968; Theng, 1974; Schnitzer and Kahn, 1972; Stevenson, 1982). We would like to emphasize only that in natural soils "pure" constituents do not exist. A mixture of 1 : 1 and 2 : 1 type clays are usually present with a predominancy towards one type or another. Colloidal oxides and hydroxides may occur in soils as a separate phase but usually as a coating or surface layer on alumino silicate clays. Organic matter and clay can form organo-clay complexes, at least up to 6% OM content—the upper limit of most irrigated soils. Herbicides interact with both organic and clay materials in such complexes (Walker and Crawford, 1968). The composition of minerals or their ratios in the mixtures is not uniformly distributed in a soil volume.

The composition of the soil solution—the liquid phase—is determined by the properties of the incoming water (rain or irrigation), as well as the characteristics of the solid phase and the residence time of the water in a particular soil volume. A decrease in the liquid–solid ratio resulting from evaporation or evapotranspiration during the wetting-drying cycles quantitatively—and sometimes qualitatively—affects the soil solution composition.

### 2. Heterogeneity

Soil transport properties on a field scale are spatially variable, sometimes even over an order of magnitude difference occurring in relatively small fields (Bresler *et al.*, 1984). Although little published data are available, it is assumed that the soils' mineralogical and organic matter content, and thus its adsorption capacity, vary as well. Beckett and Webster (1971) have calculated that the coefficient of variation for organic matter contents for a single field will be 25 to 30%. This variability is apparently inherent in all field soils to some degree, with the degree of variation an important consideration in the characterization of potential reactivity of field sites with herbicides used for crop protection in general and under irrigation in particular. Soil heterogeneity may result, for example, from microrelief effects, natural cracks, inclusion of different materials, variations imparted during formation processes, and from anthropogenic activity. Soil heterogeneity can affect herbicide behavior from the point of view of adsorption, transport, and persistence. A brief description of spatial soil variability will follow.

The interpretation of the variability of soil constituents and physical properties has been the object of interest for pedologists and soil physicists in the last few years (Biggar and Nielsen, 1976; Dagan and Bresler, 1979; Burgess and Webster, 1980; Bouma, 1981; Nielsen *et al.*, 1983; Bresler *et al.*, 1984). The concept of representative elementary volume (REV) developed and used in porous media (Hubbert, 1956; Bear, 1972) was transferred to soil science. In this concept the soil continuum is considered as a summation of representative elementary volumes (REV), and the definition and measurement of the soil properties are made on volumes of soils greater than this volume. Volume scales selected for the various soil properties differ and the REV would be different for physical, chemical, or microbiological parameters. The whole soil is considered to be divided into domains of microscopic, macroscopic, and megascopic effects, and the variations inside each domain increase from microscopic to megascopic.

When a herbicide is applied to a soil its concentration in the liquid phase is dependent not only on the soil constituents but on their domainal distribution. For example, in the microscopic domain, the concentration of the solution will fluctuate depending upon the size and distribution of the pores and of the soil particles. Solute diffusion into the particles will affect the adsorption-desorption process, the rate of chemical conversion, and the ratio between anaerobic and aerobic microbial activity. The partition between the air, water, and solid phases in the microscope domain also may affect the volatilization of herbicides. In the macroscopic or megascopic domain, where large pores or soil cracks exist, convective transport together with diffusion into and out of the soil aggregates will define herbicide behavior in soil. As a consequence the spatial distribution of soil properties affects and governs herbicide behavior in the soil system.

## C. Herbicide Application

The fate of herbicides in the soil environment may be affected by the method by which they are applied to soils. Herbicides can be applied directly to the soil in a liquid formulation or as granules, and incorporated into the soil and applied through irrigation systems. Comprehensive discussions on ground application of herbicides were recently published by Hartley and Graham Bryce (1980) and Combellack (1984).

Whatever the method of application, the herbicide is applied on the soil in the particulate form ranging from a fine spray to granules several millimeters in size. Each formulated granule of herbicide or soil particle on which pesticides are adsorbed represents a discrete source of chemicals to be redistributed in the irrigated soil by diffusion and convective flow. The heterogeneity of herbicide distribution as a result of mechanical or environmental factors is added to the heterogeneity of the soil unit, making the prediction of distribution pattern on or into the soil much more complicated. Surface application also favors loss of herbicide by volatilization and runoff, increasing the distribution variability of the materials.

Soil application of herbicides by injection or incorporation by mechanical mixing affects the distribution pattern of the chemical in irrigated soils. Soil injection of herbicides may reduce their volatilization. A deeply injected chemical is less likely to be leached downward if a low rate of irrigation is used. Mechanical incorporation reduces volatilization but increases the heterogeneity of distribution.

A herbicide may also be applied to the soil by being sprayed in the irrigation furrow before or after irrigation in flooded paddy soils, or together with the irrigation water (herbigation). The pattern of chemical incorporation under irrigation is affected by the irrigation regime, in addition to the other factors that define the behavior of a herbicide in nonirrigated soil.

The formulation of herbicides may affect their distribution in soil. Most liquid-distributed herbicides use water as a diluent. If the herbicide is not soluble in water, it is suspended in water in particulate form (wettable powder or emulsifiable concentrate). The increasing trend in "low-volume" spraying has resulted in the use of diluents other than water. Also, a herbicide may be applied to the soil as a solid diluted with dust or in granular form. Clay—mainly attapulgite—is one of the most widely used diluents for dusts and granular preparations. In both cases additional compounds are added for herbicide formulation. Some formulations contribute a slow-release effect to the herbicides. In many cases redistribution in soil of formulated herbicides following irrigation may differ from the behavior of the unformulated compound.

## D. Irrigation Technology

Irrigation technology is one of the main factors governing the fate of a herbicide in an irrigated soil. Irrigation may create saturated or unsaturated conditions and thereby define the transport of herbicides in soil. Various behavior processes of herbicides like adsorption-release, volatilization, and decomposition may be influenced by irrigation technology. Gravity, sprinkler, and drip irrigation are the main ways of conveying water from the main source to the field. Detailed information on irrigation technology may be found in specialized monographs (e.g., Jensen, 1980; Finkel, 1982). Consideration of "water distribution or redistribution in soils under various irrigation regimes"—distribution that may affect the fate of the applied herbicides—is hereby discussed.

### 1. One-Dimensional Flow

When the irrigation method leads to uniform vertical water transport we are dealing with one-dimensional flow. Surface irrigation techniques like basin or sprinkler irrigation induce a one-dimensional water flow. In discussing the water content profile under irrigation, we have to refer to infiltration and redistribution. Bresler (1971), reporting on the effect of the irrigation method on soil water regime during the infiltration stage, shows that in flooding systems the rate of water entry into the soil and the resulting water content profile are determined solely by the hydraulic properties of the soil. In sprinkler irrigation, the rate of water infiltration can be controlled within the hydraulic limit of the soil by the combination of sprinkler nozzle size, operational pressures, and sprinkler spacing, which define the sprinkler intensity. For equal amounts of applied water the wetting depth decreases and the average water content increases as the irrigation intensity becomes higher. Flood irrigation gives the highest values of water content and the shallowest depth of wetting. At the cessation of infiltration, when water supply to the soil surface stops, a redistribution of soil moisture begins. Water drains from the wet zone above the wetting front to the underlying soil. In this redistribution process the water content above the infiltration wetting front decreases while below this depth the water content increases. Thus, wetting and drying are taking place simultaneously in the different parts of the soil profile.

### 2. Two- and Three-Dimensional Flows

In several types of irrigation, water is delivered to a point or a line representing only a small fraction of the soil's total surface area. Methods inducing two-dimensional irrigation are furrow and drip irrigation, while subsurface irrigation favors a three-dimensional flow. Raats (1984) discussing the dynamics of two- and three-dimensional flow points out that once water enters the soil—as a result of application by one of the above irrigation methods—the flow is governed by gravity and capillarity just as for the one-dimensional case. However, the dynamics of the flow will be more complex with the water moving, not only in the upward and downward directions, but rather in a two- or three-dimensional pattern. For one-dimensional flow the water content increases in a limited volume of soil near the wetting front, whereas in multidimensional cases the added water is distributed over a continually increasing volume.

3. Unsaturated and Saturated Conditions

The persistence of herbicides in soil is highly affected by the microbial activity that is governed by aerobic and anaerobic conditions. Methods of irrigation may affect the soil–water–air ratio. Under some surface irrigation methods water is turned on until the desired gross volume has been applied to the area and water is ponded until infiltrated. The time of flooding—established as a function of soil hydraulic properties and crop requirement—may range from minutes or hours to months. Anaerobic conditions are formed in the upper layer of the soil profile where, in general, the herbicide is applied, which may affect its behavior. Under other surface irrigation systems (furrow, sprinkler) or drip irrigation, unsaturated conditions are predominant. A saturated zone may occur only around the furrow during the water application, but in the case of sprinkler or drip irrigation the rate of water application is usually not greater than the infiltration capacity of the soil.

### E. Fluctuation of Environmental Conditions

The main objective of irrigation is to provide plants with sufficient water to prevent stress that causes reduced yield. The required timing and amount of applied water are governed by the prevailing climatic conditions, the crop and its stage of growth, soil water holding capacity, and the extent of root development. The soil water depleted during the growing season owing to evapotranspiration is replaced by irrigation. An idealized graph of matrix potential as a function of time in an irrigated area is shown in Figure 1

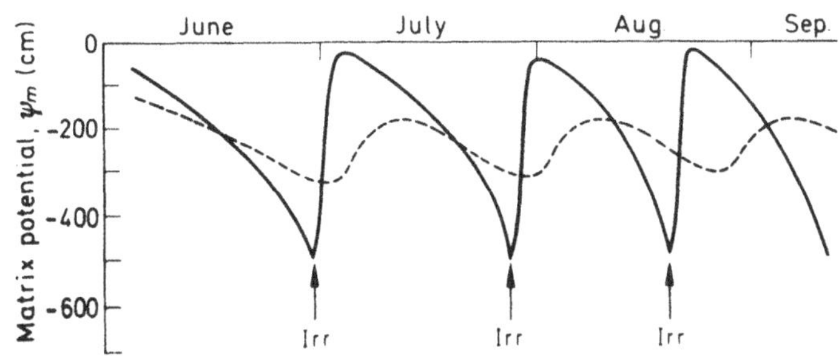


**Figure 1.** Idealized graphs of matric potential as a function of time near the top of the root zone (solid curve) and towards the bottom of the root zone (broken curve). Irrigations occurred near the end of June and July and near the middle of August. After Hanks and Ashcroft (1980).

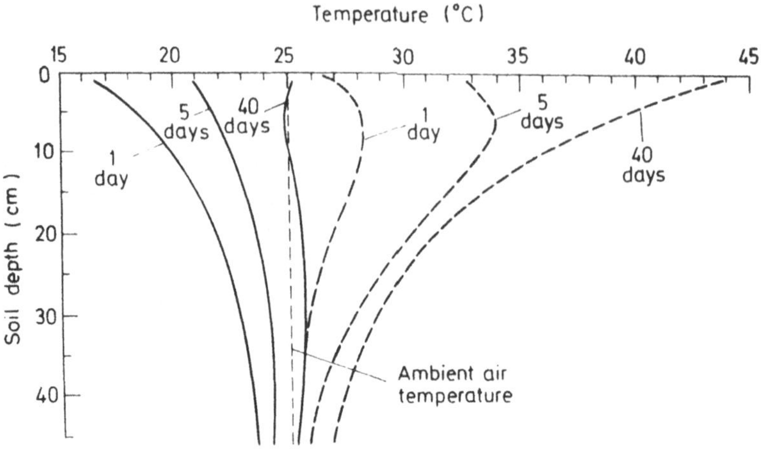


**Figure 2.** Soil temperature profiles of initially wet soil columns that were dried in a constant temperature room held at 25°C. The solid curves represent a column that was dried by blowing air across the surface. The broken curves represent a column dried by exposing it to radiation from heat lamps. After Hanks *et al.* (1967).

(Hanks and Ashcroft, 1980). It may be observed that during an irrigation season a wetting-drying cycle characterizes the soil when the gradient between wet and dry conditions is greater in the upper layer of the root zone than at the bottom.

The heat content of the soil is associated with changes in the soil water regime during the wetting-drying cycles. During the irrigation season, both the soil heat capacity and its thermal conductivity are subject to fluctuations. Figure 2 shows an example of a soil temperature profile of a wet soil during the drying process (Hanks *et al.*, 1967). Since the herbicide distribution and persistence in the soil are affected by the soil moisture content and soil temperature, it is evident that fluctuations in these parameters may change the dimension of their effect on herbicide fate.

## III. Interactions and Persistence

### A. Adsorption-Desorption on Soil Constituents

Adsorption may be defined as the excess solute concentration at the liquid–soil interface over the concentration in the bulk solution regardless of the nature of the interface region or of the interaction between the solute and the solid surface causing the excess (Mingelgrin and Gerstl, 1983). By definition, adsorption removes a compound from the bulk solution and by doing this it greatly affects the behavior of the compound in the soil environment.

The rate and extent to which a herbicide distributes itself between the solid and solution phase in a soil are determined by several physical and chemical properties of both the chemical and the soil as well as by environmental conditions.

1. Adsorption

The accepted way of expressing the extent of adsorption of non-ionic compounds is by an "adsorption coefficient" obtained from an adsorption isotherm. For dilute solutions where the mole fraction of the adsorbate is low, as is the case of herbicides in soils, the adsorption isotherm can usually be described by the Freundlich isotherm:

$$\frac{x}{m} = KC^{1/n} \qquad [1]$$

when $x/m$ is the amount of herbicide adsorbed per unit mass of soil, $C$ is the equilibrium solution concentration, and $K$ and $1/n$ empirical parameters. If the value of $n$(or $1/n$) is close to unity, a linear isotherm is obtained, and this greatly simplifies the use and comparison of $K$ values from one system to another. Values significantly different from unity pose serious problems in interpretation of results and care should be used in such cases (Bowman, 1981). Hamaker and Thompson (1972) have reported values of $1/n$ as low as 0.3 and as high as 1.7, but the general range is from 0.7 to 1.1. The adsorption coefficient for a given herbicide varies greatly from soil to soil. For example, the $K_d$ values of napropamide—measured in six soils with various organic matter and clay contents—ranged from 0.27 to 2.94 ml/g (Gerstl and Yaron, 1983a). In many cases the use of $K_{oc}$, which is defined as

$$K_{oc} = \frac{K \times 100}{\%\ \text{organic carbon}} \qquad [2]$$

will reduce the variability quite significantly. For napropamide the $K_{oc}$ values range from 249 to 450 with an average value of 336 ± 68 (Gerstl and Yaron, 1983a). The $K_{oc}$ is generally assumed to be independent of the properties of the soil or sediment, and this holds true primarily because of the similarity of the organic matter in soils throughout the world (Adams, 1972). Regardless of the exact mechanism responsible for adsorption by soil organic matter, the fact remains that the use of $K_{oc}$ is a valuable tool in assessing herbicide behavior in a soil for which actual adsorption data are unavailable. In many cases, in fact, even the $K_{oc}$ need not be available for obtaining a first estimate of a compound's behavior in soils. The $K_{oc}$ may be estimated from empirical relationships with either the compound's aqueous solubility or its octanol-water partition coefficient (Mingelgrin and Gerstl, 1983; Briggs, 1981; Hasset *et al.*, 1980). These relationships not only may be obtained for a series of similar compounds (e.g., triazine herbicides or substituted areas), but may also include a wide variety of chemicals (insecticides, herbicides, fungicides, and aromatic and polynuclear aromatic hydrocarbons) covering a range in properties of several orders of magnitude. The use of these relationships, if done with care and consideration for their limitations (Mingelgrin and Gerstl, 1983), can provide a useful value within an order of magnitude of the true adsorption coefficient. If more exact data are required, the parameter should be measured independently.

Several herbicides are ionic in nature and thus their behavior can not be understood in light of the previous discussion applicable to weakly polar chemicals. Herbicides such as paraquat and diquat have a permanent charge while other compounds become charged from processes such as protonation of weak bases (e.g., triazines) or by dissociation with the release of a proton from weak acids (e.g., picloram, 2,4-D). The primary adsorption mechanism for cationic compounds is ion-exchange with ions on the soil clay and organic matter (Burns *et al.*, 1973a; Weber and Weed, 1968). Arnold and Farmer (1979) have proposed that picloram, with a $pK_a$ of 3.6, is adsorbed primarily in the molecular form and have shown that the adsorption coefficient indeed does increase as the pH decreases.

Adsorption isotherms are measured at equilibrium, so little work has been carried out on adsorption kinetics. In batch or slurry type experiments the natural soil structure is destroyed and so the rate of adsorption on the soil surfaces will generally be rate limiting. In actual field situations molecular diffusion of the herbicide into a soil aggregate quite often becomes the rate-limiting step. This will be discussed more fully in the section on transport.

### 2. Desorption

Desorption isotherms provide information on the release of adsorbed molecules. In many cases the desorption isotherm is nearly identical to the adsorption isotherm. However, numerous examples have been reported where hysterisis in the adsorption–desorption process is quite pronounced (Savage and Wauchope, 1974; Hornsby and Davidson, 1973; Farmer and Aochi, 1974). Hysteresis may be attributed to kinetics in which the rate of desorption is unusually low or to changes occurring to the adsorbate-adsorbent complex upon continued shaking or mixing (Calvet, 1980). The existence of hysteresis is of importance for transport phenomena and for biological activity where adsorption controls the molecules available for participation in these processes. Figure 3 shows, for example, an adsorption-desorption isotherm of a herbicide.

### 3. Factors Influencing Adsorption

It is beyond the scope of this paper to review all the existing literature discussing the various factors influencing the adsorption of organic

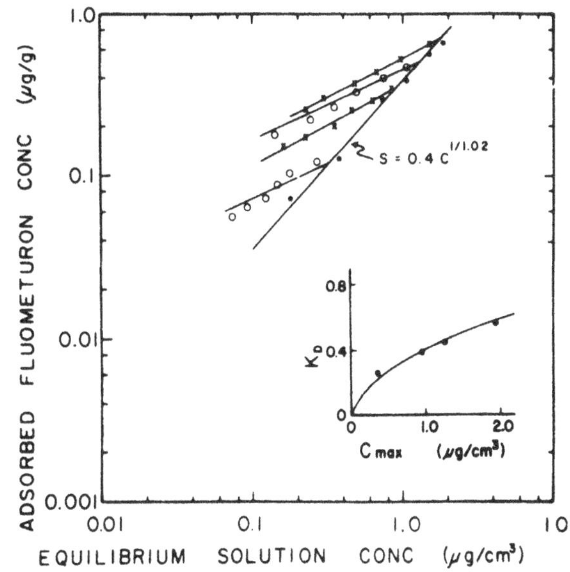


**Figure 3.** Equilibrium adsorption and desorption isotherms for fluometuron and Norge loam, a fine silty, mixed, thermic Udic Paleustolls. Insert shows relationship between the desorption distribution coefficient, $K_D$, and the maximum fluometuron solution concentration, $C_{max}$. After Davidson *et al.* (1975).

chemicals, and the reader can choose from numerous review articles (Calvet, 1980; Hamaker and Thompson, 1972). The following discussion will briefly summarize their findings.

Adsorption is determined by the complex interrelationship of (1) the molecular properties of the adsorbate, (2) the nature of the soil constituents, and (3) the experimental and environmental conditions under which adsorption is studied.

The molecular properties of greatest importance in adsorption reactions are the compounds' electronic structure and water solubility. We have already seen that ionic compounds behave differently from non-ionic compounds. The molecular structure determines the ionization potential of a compound, and this has been discussed in detail for the 1,3,5-triazines (Weber, 1967). The ability of a molecule to ionize, which depends upon the electron distribution in the molecule and on its electron mobility, is the primary reason why many herbicide–soil interactions are highly pH dependent.

For non-ionic compounds, aqueous solubility is the most important molecular parameter that affects adsorption. In general, highly soluble chemicals have relatively low adsorption coefficients for soils and low bioaccumulation factors in aquatic organisms, while the opposite holds true for chemicals of low solubility. These relationships have been shown to hold for a wide variety of compounds. In the case of herbicides, however, many exceptions have been reported. Hance (1965) did not observe any relationship between adsorption and the solubility of substituted ureas, and Weber (1967) reported that the differences in the solubility of *s*-triazines were not reflected in their adsorption. Harris and Warren (1964) concluded that no general relationship exists between solubility and adsorption of herbicides of widely different chemical groups. One possible explanation for these findings may be the effect of pH on adsorption for certain groups of herbicides or the existence of non-linear isotherms. Furthermore, recent work (Bowman and Sans, 1979) has shown that the solubilities reported in early sources are not always accurate, thus decreasing the accuracy of correlation measurements.

Formulations of herbicides affect their solubility and enable easy application of compounds with low solubilities (van Valkenburg, 1973). This does not always affect herbicide behavior in the field since most formulations readily lose their formulating powers when reaching the soil. Saltzman *et al.* (1981) have shown that no difference in behavior was noted for the technical, wettable powder and emulsifiable concentrate form of the herbicide napropamide.

The soil constituents that control adsorption are the soil organic matter, clay minerals, and amorphorus minerals. In soils the organic matter and clay minerals do not exist as separate entities, so it is difficult to know which component is responsible for adsorption. Many studies have dealt with "simple" systems that contain only one adsorbent (Gerstl and Yaron, 1977; Hayes, 1970; Khan, 1979) while other studies have used actual field soils and related adsorption to the clay or organic matter content (Felsot and Dahm, 1979; Gerstl and Yaron, 1983a; Hance, 1965). From both types of studies, much valuable information can be obtained, and the following discussion will deal with each component separately.

Clay minerals are made up of sheets of silica tetrahedra and alumina octahedra and occur in soils in hydrated forms. Clay minerals have a negative charge that is balanced by exchangeable cations and thus are important in adsorption of cationic and protonated herbicides. Paraquat and diquat adsorption by the external surfaces of vermiculite and mica are linearly related to the surface charge densities (Philen *et al.*, 1970, 1971). The pH in the immediate vicinity of a clay surface is much lower than in the bulk solution. Consequently, clays can be very strong proton donors and as a result many herbicides will be adsorbed in the protonated form by clay minerals (Bailey *et al.*, 1968). At very low pH values (2–5) the clay edges may retain a net positive charge, thus favoring adsorption of anionic compounds such as picloram and 2,4-D.

Soil organic matter (OM) is a complex polymeric mixture arising from microbial and chemical degradation processes, the exact structure of which has not yet been fully elucidated. Soil OM is generally held responsible for the adsorption capacity of soils for non-ionic pesticides, and this is borne out by the constancy of $K_{oc}$ values for a given compound in a wide variety of soils. In soils with very low organic matter contents (<0.1% OM), the $K_{oc}$ values are usually much greater since adsorption by other soil constituents is being attributed to OM (Gerstl and Mingelgrin, 1984; Hasset *et al.*, 1980). The water-insoluble humic acids and humins that are important constituents of soil OM have been reported to be responsible for most of the adsorption of organic compounds in soils (Burns *et al.*, 1973a, 1973b) while non-humic substances contribute little to the adsorption of organic compounds.

The surfaces of clay minerals are generally hydrophilic but by exchanging the inorganic cation with an organic cation the nature of the surface can be greatly altered. Gerstl and Mingelgrin (1979) have shown that the adsorption capacity of the clay surface for pesticides was greatly increased on attapulgite clay adsorbed with trimethyl ammonium or hexadecyltrimethyl ammonium. Processes similar to these probably occur in soils where clay-organic matter complexes exist. Hance (1969) claimed that the presence of such clay-OM complexes in soils reduced the adsorptive capacity of the individual components.

Amorphous materials can predominate over clays and OM in certain soils (Adams, 1973). These amorphous materials are oxides and hydroxides of Fe and Al and allophane and their adsorptive capacities are all affected by the soil pH. Hilton and Yuen (1963) reported that the amorphous materials in Hawaiian soils greatly influenced the adsorption of pre-emergence herbicides. Hamaker *et al.* (1966) found that Fe and Al sesquioxides adsorbed picloram more strongly than clays. Wauchope (1975) found a strong correlation between the iron-oxide content of soils and the adsorption of organoarsenical herbicides, and O'Connor and Anderson (1974) found similar results for 2,4,5-T.

Very little work has actually been done to study the efficacy of herbicides in the field as related to their adsorption in a wide variety of soils. Cussans *et al.* (1982) applied chlorotoluron and isoproturon to 44 fields throughout England and kept records on the degree of weed control. Soil samples were also taken and the $K_d$ values of chlorotoluron determined. Their results show that $K_d$ values seemed to have greater predictive potential for herbicide efficiency in the field than did organic matter analysis. For example, 9% of the fields with $K_d$ values above 6.0 exhibited poor weed control, while 78% of fields with $K_d$ values between 4.5 and 6.0 exhibited variable control.

The study of adsorption by laboratory methods, whether it be by batch, column, or another method, may not adequately describe the adsorption process in the field. The aggregation of soil particles in the field will affect the exposure of adsorptive surfaces for organic compound adsorption, and attainment of adsorption equilibrium may be very slow since access to the adsorptive sites within the aggregate will be diffusion controlled. If a compound is leached through a soil at a rate greater than its diffusion into (or out of) a soil aggregate, equilibrium will not be reached, and the compound will move deeper into the soil than would otherwise be expected from laboratory adsorption data. White *et al.*[1] reported bromacil and napropamide were found in the drainage water of a 25 cm undisturbed core of an Oxfordshire soil with 15% organic matter ($K_d$ of napropamide was 24.4 ml/g) when applied in the irrigation water and leached immediately owing to bypass flow in macropores. If, on the other hand, leaching was delayed for 24 hrs the effluent concentrations were much lower because the compounds were able to diffuse into the aggregates, thus becoming unavailable for leaching.

The final factors influencing adsorption are those of the experimental conditions such as temperature, pH, ionic composition, and soil:solution ratio.

Changes in temperature can either increase, decrease, or leave unchanged the extent of adsorption in a given system (Gerstl and Mingelgrin, 1979; Calvet, 1980) as can be seen from Table 1 based on published results. It has generally been assumed that increasing temperature results in decreased adsorption because of an increase in the solubility of most compounds at higher temperatures.

For certain herbicides, namely weak acids and weak bases, pH has a tremendous effect on adsorption, especially since the pH of colloidal surfaces in soils is usually several pH units lower than that of bulk solution (Bailey *et al.*, 1968). In general, the neutral species of an acid adsorbs much more strongly than the anion. Weak bases are converted to the cationic form at low pH values and are more highly adsorbed than the free base. For such compounds adsorption will increase with decreasing pH as can be seen in Figure 4 for the adsorption of picloram by soils (Hamaker *et al.*, 1966). Also shown in this figure is the adsorption of 2,4-D by geothite in which a slightly different pattern may be discerned (Watson *et al.*, 1973). Decreasing adsorption with increasing pH is due to electronic repulsion because both 2,4-D and geothite are negatively charged. As the pH decreases the geothite surface become negatively charged, allowing greater anion adsorption.

The presence of salts in the soil solution can cause a decrease in the adsorption of cationic species owing to competition for exchange sites. Neutral molecules are generally less affected by salts but may show increased adsorption with increasing salt concentration, possibly because of a salting-out effect (Table 2).

Most of the material presented in this section is based upon studies carried out in suspensions. The actual conditions under which herbicides are required to function are quite different and consequently it should not be surprising if field and laboratory studies do not always agree. It is

[1]Personal communication, 1984.

**Table 1.** Effect of the Temperature on the Adsorption of Herbicides by Organic Materials, Minerals, and Soils[a]

| Adsorbent | Herbicide | Effect of temperature |
|---|---|---|
| Montmorillonite (pH 8,5) | Simazine | –[b] |
| Illite | 2,4-D | – |
| Montmorillonite | 2,4-D | 0 |
| Montmorillonite Na | Paraquat | 0 |
| Vermiculite Na | Paraquat | + |
| Peat | Monuron | 0 |
| | Simazine | 0 |
| | Atrazine | 0 |
| | 2,4-D | 0 |
| Humic acid | Atrazine | + |
| Humic acid Lignin | Atrazine | + |
| Humic acid | Atrazine | + |
| Charcoal | Prometon | 0 |
| | 2,4-D | – |
| Soils | 1,3,5,-triazines | – |
| Soils | Alachlor | – |
| Soils | Picloram | – |
| Soils | 12 dinitroanilines | – |

[a]After Calvet (1980).
[b]0 = no effect; – = adsorption decreases as temperature increases; + = adsorption increases with temperature.

necessary, therefore, that the adsorption of organic compounds under field conditions and its relationship to field movement and biological activity receive further attention.

## B. Biodegradation

Biodegradation of organic molecules in the soil is one of the most important of all environmental processes and can result in the eventual mineralization of these compounds. Ideally any herbicide used should persist long enough to give adequate weed control without having serious carryover effects. Like most processes occurring in nature, the persistence, or biodegradation, of a particular compound is dependent upon numerous chemical and environmental factors that differ from one site to another in any given field and that can change greatly with time. This limits, to a great extent, the results of field trials to the particular location, season, and compound under study and so most work done on biodegradation and factors that affect this

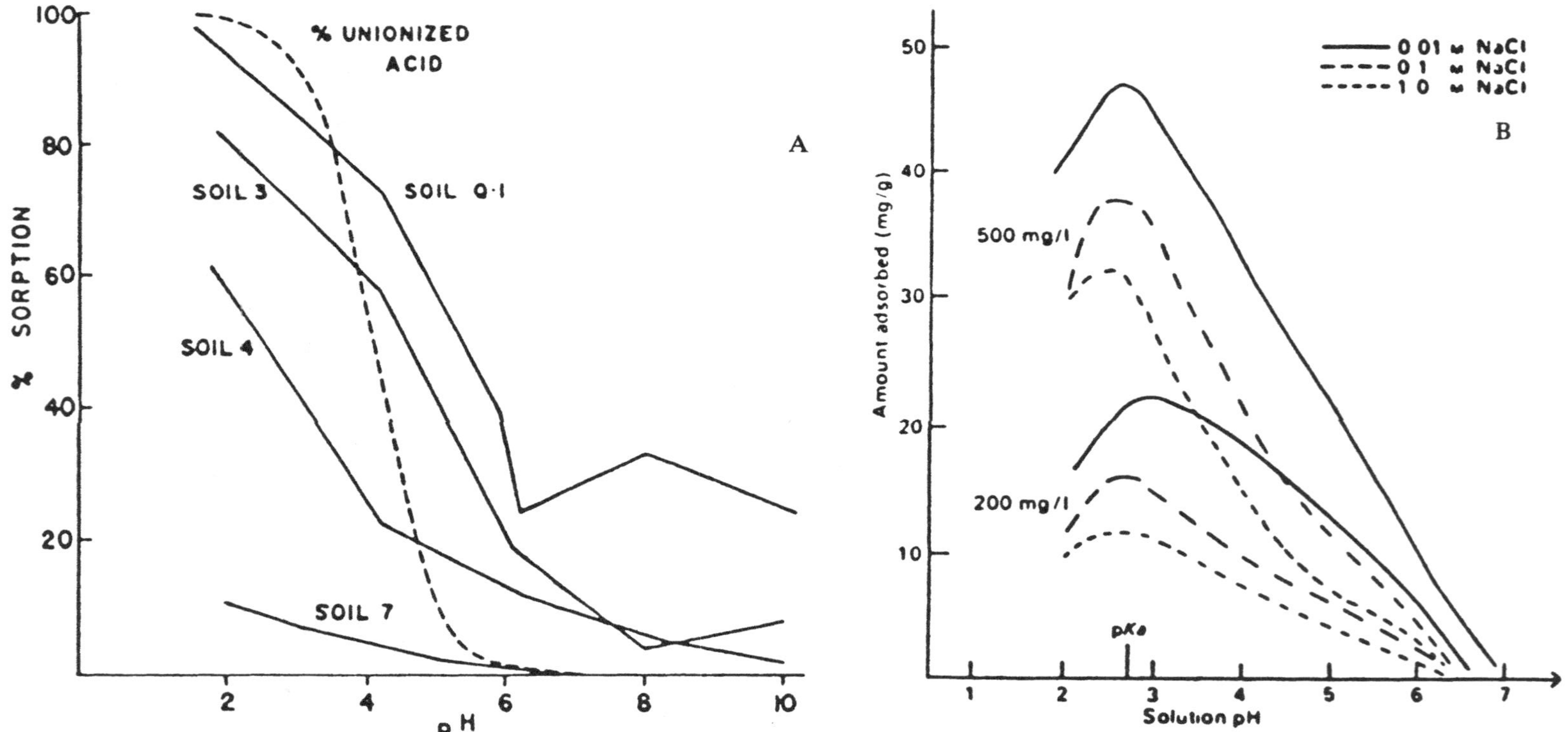


**Figure 4.** Effect of the adsorption of acidic herbicides by soil components. (A) Adsorption of picloram by soils. Reprinted with permission from Hamaker *et al.*, *Adv. Chem. Ser.* 60: 23–37. Copyright 1966, American Chemical Society. (B) Adsorption of 2,4-D on goethite as affected by pH and ionic strength. From Watson *et al.* (1973).

**Table 2.** Effects of Salts on Adsorption of Herbicides by Soils and Soil Components[a]

| Adsorbent | Herbicide | Salt effect: Ionic strength <1 | Salt effect: Ionic strength >1 | Experimental conditions | |
|---|---|---|---|---|---|
| Montmorillonite Na | Fenuron | –[b] | + | NaCl | |
| Montmorillonite Na | Fenuron | 0 or – | + | NaCl | |
| Montmorillonite Mg | Monuron | | | $MgCl_2$ | Identical concentrations |
| Montmorillonite Ca | | | | $CaCl_2$ | |
| Soils | Terbutryne | – | + | NaCl | Identical ionic strength |
| | Atrazine | | | KCl | |
| | | | | $CaCl_2$ | |
| Goethite | 2,4-D | – | + | NaCl | |
| Soils | 1,3,5-triazines | + | | KCl | Identical ionic strength |
| | Substituted ureas | | | $NH_4Cl$ | |
| | | | | $CaCl_2$ | |
| Soils | Picloram | + | | KCl | Identical concentrations |
| | | | | $CaCl_2$ | |
| Soils | **Prometryn** | + | | NaCl | Identical concentrations |
| | Fluometuron | – | | $CaCl_2$ | |

[a] After Calvet (1980).
[b] 0 = no effect; – = the adsorption decreases as the ionic strength increases; + = the adsorption increases with the ionic strength.

important process have been carried out under controlled conditions in the laboratory.

Biodegradation may be defined as any biologically induced structural transformation in the parent compound that changes its molecular integrity (Scow, 1982). The vast majority of the reactions involved in the biodegradation of herbicides may be classified as oxidative, reductive, hydrolytic, or conjugative (Hill, 1978). This section will not deal with individual compounds or processes but will briefly review those findings that deal with the factors governing these reactions.

### 1. Chemical and Environmental Factors Affecting Biodegradation

#### a. Soil Type

The effect of soil type on herbicide persistence can not be easily defined. On one hand, soils with higher organic matter content tend to have greater microbial activity but at the same time tend to adsorb the herbicide more strongly, thus reducing its soil solution concentration, possibly protecting it from degradation. On the other hand, the distribution of microorganisms in soils is not uniform, their density usually increasing near colloidal surfaces (Burns, 1978). Furthermore, several investigators report (Hance, 1970; Russel *et al.*, 1968; Probst and Tepe, 1969) that adsorbed compounds may undergo non-biological catalyzed reactions resulting in more rapid herbicide losses.

Hamaker (1972a) postulated that in mineral soils the rate of degradation will increase up to a certain organic matter content after which no increase in degradation rate will be noted. The data from the literature both supports and conflicts with this suggestion. Support comes from studies with linuron, simazine, 2,4-D, TCA, and other herbicides for which the rate of decomposition was greater in soils with higher organic matter contents (Walker and Thompson, 1977; Zuarawski and Piss, 1971; Ogle and Warren, 1954). However, Harris *et al.* (1969) found that as soil organic matter increased so did the persistence of fenac. The results of Davison and Clay (1972), Meikle *et al.* (1973), and Hamaker *et al.* (1967) further complicate matters because they found no relationship between herbicide (fenac and picloram) degradation and any soil property.

One possible explanation for these conflicting results is that other factors played a role in determining the microbial activity in these soils. A case in point can be the presence or absence of microsites in soils which is, to a certain extent, dependent upon soil structure and agrotechnical practices (Haider, 1983). Haider (1983) points out that adsorption of microbes by soil colloids as well as local differences in pH and nutrients at soil surfaces can result in microsites in which the center of soil aggregates can be oxygen-depleted although their surface is exposed to adequate oxygen concentrations. As a result, reactions such as nitrification and denitrification can occur simultaneously. In respect to herbicides, trifluralin and benefin, two dinitroaniline herbicides, were found to undergo reduction and/or dealkylation in both anaerobic and aerobic soils (Williams, 1977; Probst *et al.*, 1975).

#### b. Soil pH

The same difficulties encountered in trying to assess the effect of soil type on biodegradation exist when it comes to defining the effect of soil pH on the rate and extent of herbicide decomposition in soils. In addition to the reasons stated in the previous section, the soil pH can also affect degradation through its effect on the chemical's stablity, or by its effect on adsorption, or the makeup of the soil microflora. The artificial adjusting of soil pH in degradation studies is rarely justified since this treatment may change other soil characteristics or cause drifts in the predominant form of microorganism (Hamaker, 1972a).

Hurle and Walker (1980) have recently reviewed the available literature on the effect of pH on biodegradation, and from their review it is obvious that no general conclusion can be made. For instance atrazine degradation was found to increase as the soil pH decreased (Best and Weber, 1979), decrease as the soil pH decreased (Hance, 1979) or be nearly unaffected (Hance, 1979; Dao *et al.*, 1979)

Because of the complexity of response to pH its effect is largely unpredictable and should be determined experimentally whenever required.

#### c. Temperature and Moisture Content

Environmental variables such as temperature and moisture content do not control biodegradation specifically; rather, they control microbial metabolic activity in general. Most microbial processes in soils are mediated by mesophilic organisms that have an optimal range of 25 to 40°C (Brock, 1970). In soils, especially, temperature is interrelated to other environmental parameters. Moist soils conduct heat more efficiently so that temperature gradients over a given depth are smaller than in dry soils (Raney, 1965). In general, it has been found (Hamaker, 1972a) that the rate of herbicide degradation increases with increasing temperature.

Moisture content is important because it controls the oxygen level in soils by competing for pore space. When only 10 to 20% of the total pore space is gas filled, anaerobic conditions form (Brock, 1970). Water is also required by microorganisms to prevent rupture of their cell membranes (Gray, 1970), although some species can form endospores and wait for more favorable conditions. Most bacteria can tolerate soil water potentials of 5 bars while some can survive at potentials as low as 80 bars.

Much effort has gone into studying biodegradation. However, the number of studies in which several temperatures and moisture contents were employed is fairly limited. Parker and Doxtrader (1983) studied 2,4-D decomposition and observed a lag period in which only little degradation occurred and after which a rapid phase of decomposition began. The length of this lag period increased with decreasing moisture content but disappeared altogether at temperatures above 37°C. The degradation of picloram in five soils under drying conditions was reported by Guenzi and Beard (1976). Little effect of temperature was noted between 5 and 25°C with a sharp increase in decomposition rate at 30°C. Degradation decreased as the water content decreased and ceased when the soil was air dried. Alternating wetting and drying cycles imposed on several soils led to a decrease in decomposition following each successive drying cycle. These results augur profound effects with differing irrigation schemes, for instance drip vs. sprinkler where in the former the soil is kept at a more constant and greater moisture content than in the latter. Gerstl and Yaron (1983a) studied the behavior of bromacil and napropamide in several soils at several temperature and moisture contents and found that soil type and initial concentration were less important for degradation than environmental conditions. They found that the half-life of napropamide in two soils kept at field capacity decreased with increasing temperature up to 35°C. Above 35°C (up to 50°C) no temperature effect was noted. For bromacil half-lives decreased with increasing temperature up to 50°C. For both herbicides the half-life decreased significantly as moisture content increased (from 30 to 100% of field capacity) when incubated at 25°C.

A model proposed by Walker (1974) that quantitates the effects of temperature and moisture content on herbicide degradation will be discussed in a separate section.

#### d. Other Factors

Many other factors affect the rate of herbicide loss in soils. The initial concentration of a herbicide has been shown to affect degradation kinetics. Reduced degradation rates at higher initial concentrations have been attributed to possible toxic effects on microorganisms or enzyme inhibition (Hurle, 1982). Edwards (1972) claims that in the field only unusually high rates of herbicide application would produce such an effect.

Since biodegradation of most herbicides is controlled to a large extent by microbial activity, the addition of a readily available energy source, which will result in a nearly instantaneous increase in microbial activity, is expected to result in enhanced herbicide degradation. Once again as with many other "expectations" many exceptions are to be found.

The addition of glucose, nutrient broths, ground vetch tops, and corn and bean straw resulted in the enhanced degradation of many herbicides (McClure, 1970; Savage, 1977; McCormick and Hiltbold, 1966; Wolf and Martin, 1974). Doyle *et al.* (1978) observed an increase in microbial activity when sewage sludge and dairy manure was applied to soils, but the former inhibited degradation while the latter increased it. Hance (1973) also reported on a case of farmyard manure increasing degradation of atrazine in one soil but not in another, while manure, fertilizers, or both in combination did not affect linuron decomposition. To fully understand the effect of soil amendments on biodegradation, information is needed on the specific interaction between herbicides and microbes—information that is more often than not lacking.

## 2. Enhanced Degradation

The problem of enhanced degradation that is accompanied by an excessively rapid loss in efficacy of applied pesticides poses a serious problem in the fight to control crop pests (Harris *et al.*, 1984). Recent laboratory and field studies have shown that the interaction of herbicides (and other soil-applied pesticides as well) with microorganisms has resulted in many cases of reduced persistence and efficacy (Kaufmann *et al.*, in press). This phenomenon is not limited only to the chemical in question, but affects other chemically related compounds. Several workers have studied this phenomenon for the herbicides butylate and EPTC (Miaullis *et al.*, 1982; Rudyanski and Fawcett, 1983; Menkveld and Dekker, 1983; Tuxhorn *et al.*, 1983). They observed that prior butylate application affected its persistence and weed control ability. Butylate plus R-25788 (Butylate$^+$) and EPTC$^+$ lost all weed control 3 and 5 weeks after application to soils that had received butylate several years before, while in sterile soils both herbicides remained effective. In other cases, use of R-33865 instead of R-25788 significantly increased persistence of the herbicides. The similarity of R-33865 to many organophosphate insecticides leads one to wonder if certain herbicide–insecticide combinations may not be a worthwhile topic for research. The data reemphasizes the importance of soil microbial activity and its role in the degradation of these herbicides. Further research is needed to evaluate the importance of this microbial adaption to the rapid disappearance of herbicidal activity because it has important implications for crop protection practices.

## 3. Bound Residues and Carryover Effects

Biodegradation experiments are generally carried out by extracting the remaining intact pesticide or by measuring the $^{14}CO_2$ released from labeled material. Since the early 1970's it has become apparent that not all of a given compound present in a soil is extractable by conventional methods (Chiska and Kearney, 1970; Bartha, 1971). These residues have been termed "bound residues" (BR) and have been defined as non-extractable (bound) pesticide residues in soils, plants, and foods—originating from pesticide usage that cannot be extracted by methods commonly used in residue analysis and metabolism studies (Hassan, 1982). Examples of bound residues of herbicides are presented in Table 3 taken from Khan

**Table 3.** Bound Residue Levels of Some Herbicides in Soils[a]

| Pesticide | Soil | Organic matter (%) | Rate (ppm) | Time[b] | Bound $^{14}C$ residues (% of applied) |
|---|---|---|---|---|---|
| Butralin | Silt loam | 1.5 | 10 | 7 m | 17 |
| Chlornidine | Silt loam | 1.5 | 10 | 7 m | 17 |
| 2,4-D | Sandy loam | 4.0 | 2 | 35 d | 28 |
| 3,4-Dichloroaniline | Sandy loam | 1.0 | 0.7 | 18 m | 90 |
| Dinitramine | Silt loam | 1.5 | 10 | 5 m | 8 |
| | Silt loam | – | 0.6 | 244 d | 55 |
| Flampropisopropyl | Peat | – | 50 | 28 w | 19 |
| | Sandy loam | – | 17 | 28 w | 10 |
| Fluchloralin | Silt loam | 1.5 | 10 | 7 m | 21 |
| Isopropalin | Silt loam | – | 11 | 12 m | 27 |
| Methabenzthiazuron | Sandy loam | 2.2 | 10 | 111 d | 41 |
| Pirimicarb | – | – | 0.5 | 24 m | 70 |
| Oryzalin | Silt loam | – | – | 36 m | 35 |
| | Silt loam | – | – | 12 m | 56 |
| Oxadiazon | Loam | 1.5 | 10 | 25 w | 13 |
| Pentachlorophenol | Silty clay loam | 2.3 | 10 | 24 d | 45 |
| Profluralin | Silt loam | 1.5 | 10 | 7 m | 11 |
| Prometryn | Organic | 85 | 12.4 | 150 d | 43 |
| Propanil | Sandy loam | 6.0 | 5 | 20 d | 73 |
| | Clay loam | 4.1 | 6 | 25 d | 73 |
| Trifluralin | Silt loam | – | 0.8 | 12 m | 50 |
| | Silt loam | 1.5 | 10 | 7 m | 7 |
| | Loam | – | – | 36 m | 38 |
| | Silty clay loam | 3.9 | 10 | 63 d | 72 |

[a] After Khan (1980).
[b] d = day; w = week; m = month.

(1980). These residues would be unimportant if we could be sure that their bioavailability were insignificant.

The kinetics of bound residues formation is shown in Figure 5 where it becomes obvious that after 150 days the process is still continuing (Khan and Hamilton, 1980). Bound residues may be the unaltered parent compound or other compounds derived from it. For example, Khan and Hamilton (1980) found that after 1 yr more than one-half of the total bound residues in a prometryn-treated soil was present as the intact parent compound. In the case of propanil, however, Bartha (1971) showed the existence of a humus–3,4-dichloroaniline complex in addition to the parent compound.

The question of bioavailability and pollution by BR's is more than

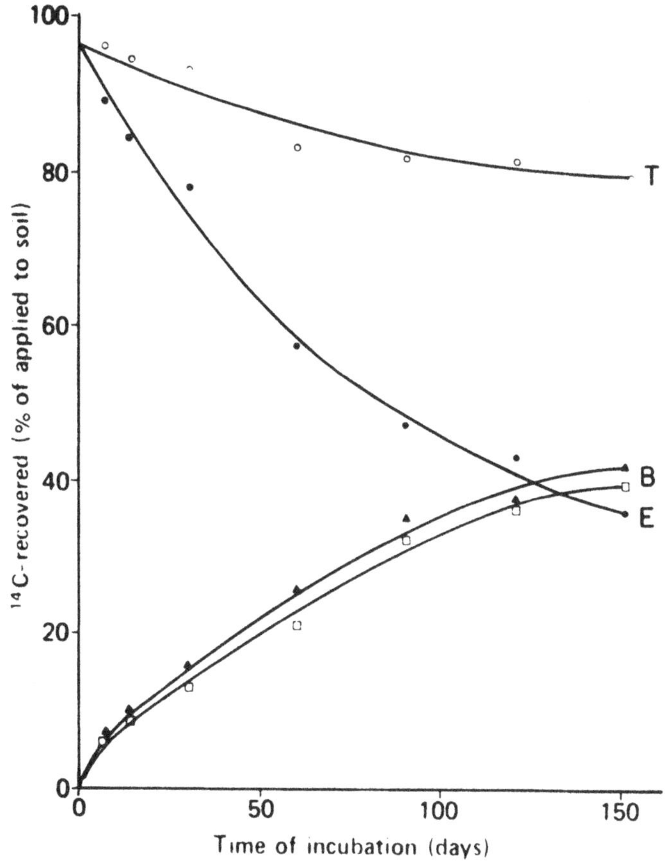


**Figure 5.** Extractable and bound radioactivity in an organic soil treated with $^{14}C$-prometryn and incubated aerobically under moist conditions for 150 days: E, extracted $^{14}C$; B, bound $^{14}C$ determined by combusting soil to $^{14}CO_2$; and T, total of extractable and bound. Reprinted with permission from Khan and Hamilton, *J. Agr. Food Chem.* 28:126–132. Copyright 1980, American Chemical Society.

academic. Twenty-six percent of the BR of prometryn found were in the fulvic acid fraction of the soil organic matter. This fraction is relatively mobile under some field conditions and can reach the ground water by the mechanisms outlined by Vinten *et al.* (1983). Khan and Ivanson (1981) have demonstrated that incubation of prometryn BR's inoculated from a control soil released significant amounts of intact herbicide. Helling and Krivonak (1978) observed that soybean (*Glycine max*) plants grown in soil containing dinitroaniline BR contained up to 1% of the $^{14}C$ residues with most of it remaining in the roots but nearly 25% reached the green tops. Fuhr and Mittelstaedt (1980) reported similar results for maize (*Zea mays*) grown on soils containing BR of methabenzthiazuron.

Until more data become available, it is difficult to say for sure whether or not bound pesticide residues will become a serious environmental problem.

Evidence so far indicates that residues will slowly be mineralized to $CO_2$ preventing the buildup of significant amounts (Gerstl and Helling, in press). If significant amounts of herbicide residues do accrue and become available for plant uptake either as a result of microbial activity or some agronomic practice (liming, manuring, etc.), serious carryover effects may result. Many cases of carryover effects have been reported by Sheets and Harris (1965) although none of them have been positively attributed to the bioavailability of bound residues.

The problem of herbicide residues (carryover effects) has recently been studied in detail by Caverly (1978) who incorporated trifluralin, linuron, and lenacil at various rates into soil at three different field sites and sowed 16 crops at each location. Soil samples were collected for residue analysis when symptoms of phytotoxicity became evident. His results (Table 4) will allow better interpretation of residue analysis for these herbicide–crop combinations. More recent work on the carryover effects of chlorsulfuron (Peterson and Arnold, 1983; Zimdahl and Fithian, 1984) has shown that even 2 yrs after application injury to crops was still evident. For certain crops, such as sugarbeet (*Beta vulgaris*), maize, and sunflowers (*Helian-*

**Table 4.** Minimum Herbicide Residue Causing Visible Symptoms of Phytotoxicity to Various Crops[a]

| | | Herbicide residue (mg/kg in 0 to 15 cm) | | |
|---|---|---|---|---|
| Crop | Botanical name | Lenacil | Linuron | Trifluralin |
| Turnip | *Brassica rapa* | 0.04 | 0.25 | 0.43 |
| Lettuce | *Lactuca sativa* | 0.07 | 0.28 | 1.06[b] |
| Wheat | *Triticum aestivum* | 0.09[b] | 0.24[b] | 0.19 |
| Barley | *Hordeum vulgare* | 0.10[c] | 0.39[b] | 0.14 |
| Oat | *Avena sativa* | 0.10[c] | 0.33[b] | 0.11[b] |
| Rape | *Brassica napus* | 0.11 | 0.24 | 0.95[b] |
| Pea | *Pisum sativum* | 0.13[b] | 0.91 | 0.82 |
| Ryegrass | *Lolium perenne* | 0.18[b] | 0.49 | 0.10 |
| Onion | *Allium cepa* | 0.18[b] | 0.77[b] | 0.66 |
| Maize | *Zea mays* | 0.18[c] | 0.41[c] | 0.12 |
| Carrot | *Daucus carota* | 0.18 | 0.91 | 2.45 |
| Parsnip | *Pastinaca sativa* | 0.18 | 0.91 | 2.45 |
| Potato | *Solanum tuberosum* | 0.21 | 0.67 | 0.82 |
| Dwarf bean | *Phaseolus vulgaris* | 0.23 | 0.38 | 0.82 |
| Red beet | *Beta vulgaris* | 0.83 | 0.49 | 0.10 |
| Flax | *Linum usitatissimum* | 0.94 | 1.00 | 1.72 |

[a]Mean of three sites, recalculated from Caverley (1978) by H. Hurle and A. Walker (1980).
[b]Data from two sites.
[c]Data from one site only.

*thus annuus*), 2 yrs after chlorsulfuron application at recommended rates crop yields were reduced by as much as 70 to 80%. Similar work has recently been reported by Walker and Brown (1982) for several other herbicide–crop combinations.

## 4. Kinetics and Modeling

Organic chemicals that undergo biodegradation reactions are generally classified into one of the three groups depicted in Figure 6. The first group that undergoes immediate degradation includes simple sugars and compounds in the proper form to enter the metabolic pathways. The second group requires acclimation, a lag period, before significant decomposition takes place. The third group of recalcitrant organic compounds are typified by substances such as humus, lignin, and several organochlorine pesticides.

Before the rate of degradation can be quantified and rate constants calculated, the correct kinetic expression must be derived to describe the pattern of loss with time. Different rate equations may be necessary for different systems owing to the complexity of factors interacting on the herbicide–soil system. In fact, one rate law may not adequately describe a herbicide's behavior over the entire time period studied; however, it is generally assumed that only one rate law applies.

Most degradation results for herbicides in soil systems assume first-order kinetics, the equation for which is $C = C_0^{e^{-kt}}$ where $C$ is the concentration at time $t$, $C_0$ is the initial concentration, and $k$ is the rate constant. Although most published results fit first-order kinetics, no mechanistic implications should be inferred, because the reactions occurring are microbially

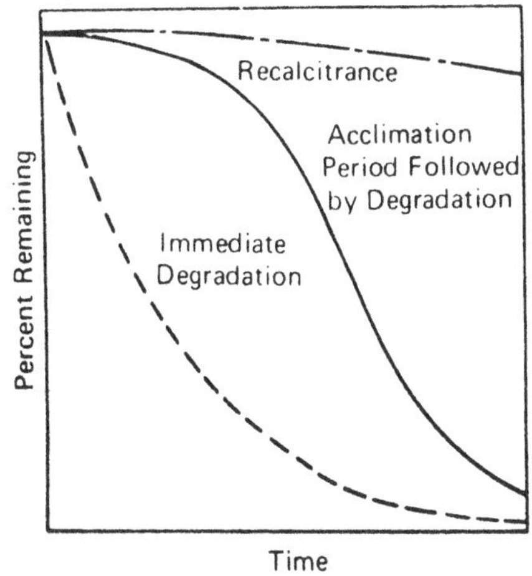


**Figure 6.** Generalized disappearance curves for organic compounds in soils. From Scow (1983).

mediated and not purely chemical. This becomes obvious when one considers the half-life (time required for 50% degradation), which for purely chemical first-order reactions is independent of the initial concentration. Many cases of herbicide degradation show a marked effect of initial concentration (Hurle and Walker, 1980). The presence of a lag period, which may vary from a few hours to several weeks, is also proof that chemistry is not working alone in the soil. Lag periods may arise from (1) the initial phase of exponential growth of one species in a mixed population that is capable of assimilating the herbicide in question; or (2) the adaption of already existing microorganisms through enzyme induction. Herbicides exhibiting lag periods include 2,4-D, MCPA, dichlorprop, DNOC, dalapon, TCA, chloridazon, propham, chlorpropham, and endothal (Hurle and Walker, 1980).

Not all cases of biodegradation are easily characterized. Hance and McKone (1971) found that for atrazine, picloram, and linuron degradation neither zero-order, half-order, first-order, nor Michaelis–Mentin kinetics described the loss pattern adequately.

The use of laboratory-derived kinetic data for predicting the loss of a herbicide under field conditions would seem to be risky at best, considering the many variables that influence biodegradation. However, Walker and co-workers (Walker, 1974; Walker and Barnes, 1981; Walker and Zimdahl, 1981; Walker *et al.*, 1983) have had considerable success with their model.

Walker's model of herbicide persistence combines laboratory measurements of the effects of soil moisture and temperature on the degradation of a compound with actual field measurements of the fluctuations of these variables in the upper 10 cm of a field soil as well as with rainfall and evaporation. Results from his field trials with numerous herbicides were generally good, even though they are based on first-order kinetics, which is only an approximation in many cases to the true kinetics. Recent collaborative studies (Walker and Zimdahl, 1981; Walker *et al.*, 1983) have shown its usefulness in a wide variety of soils and climatic conditions.

A major limitation to his model is the lack of a relationship between the experimentally derived parameters and soil or chemical properties. The model assumes the temperature effect can be described by the Arrhenius relationship that has been shown to be quite applicable to such systems. However, the two derived parameters for the effect of moisture content arise from a relationship that is entirely empirical and must be determined for every soil–herbicide combination. The model in its present state also ignores volatilization, photochemical degradation, and plant uptake of the herbicides. Despite these limitations its success has made it the best available model for herbicide persistence in field situations.

### C. Chemical Degradation in Soils

Much effort has been spent in distinguishing between chemical and microbial degradation reactions in soils. In general, the experimental approach is to use a natural and a "sterile" soil and to compare the results of transformations occurring in both. Unfortunately, however, each sterilization method has its own limitations and shortcomings. Heat and radiation sterilization methods destroy extracellular enzymes and alter the chemical and physical properties of the soil, while chemical sterilants do not usually destroy the entire soil microorganism population. Consequently, it is sometimes difficult to distinguish between purely chemical reactions and microbially governed reactions especially since in many cases there are striking similarities between mechanisms and/or degradation products.

Goring *et al.* (1975) has classified chemical transformations occurring in soils as hydrolytic nucleophilic transformations, non-hydrolytic nucleophilic displacement reactions, and oxidations.

### 1. Hydrolysis

Water is probably the most ubiquitous soil constituent and can be found in relatively high molar quantities even in apparently dry soils, thus its reaction with organic molecules would not be surprising. The term hydrolysis is applied to a reaction that involves the dissociation of water into $H^+$ and $OH^-$ by the action of substances that will combine with either species. This process is usually distinguished from several other possible reactions between water and organic compounds such as acid:base reactions, hydration, addition, and elimination reactions. It is difficult, sometimes, to distinguish between these reactions and so in this section we will consider any reaction in which water plays a role, including hydrolysis.

Hydrolysis is by far the most important reaction that organic compounds undergo in the presence of water. Compounds as diverse as alkyl halides, carboxylic acid esters, carbamates, organophosphates, and nitriles undergo hydrolysis (Harris, 1982) with half-lives ranging from several minutes for phosphoric acid esters to several hundred years for amides and phosphonic acid esters (Mabey and Mill, 1978).

The presence of catalytic surfaces in the soil can help bring about hydrolysis reactions. Armstrong and Chesters (1968) have shown that the hydrolysis of atrazine in the presence of moist clay is adsorption catalyzed. Similar reactions have been noted for other triazine herbicides (Russell *et al.*, 1968). Wet soil has also been found to bring about additional reactions.

The surface pH of clay minerals is several pH units lower than that of the bulk suspension (Bailey *et al.*, 1968). This effect has great significance in the case of triazine herbicide hydrolysis. Triazines can be hydrolyzed to their hydroxy forms in strong acids, bases, or soil suspensions (even sterile soil suspensions). Several workers have proposed that in soils montmorillonite surfaces are responsible for catalyzing the reaction (Russell *et al.*, 1968; Brown and White, 1969). Chlorotriazine becomes protonated on a ring and/or side chain N causing the chlorine-bearing ring carbon to become electron deficient enough that the OH group from water (or dissociated water) is sufficiently nucleophilic to displace the chloride ion. Similar mechanisms have been proposed for trifluralin and benefin (Probst *et al.*, 1967). Other workers (Armstrong and Konrad, 1974; Skipper *et al.*, 1967) postulated a similar sorption-catalyzed hydrolysis involving the carboxyl groups of organic matter. Stevenson (1972) claims that triazines may also be hydrolyzed by fulvic acids in the soil leading to products that are more strongly bound to the soil organic matter (bound residues). Liu and Felbeck (1972) proposed that the reactivity of atrazine with humic acid is due to steric matching.

### 2. Oxidation and Non-hydrolytic Reactions

Oxidation of pesticides in soils is primarily microbially induced and not a truly chemical reaction. Khan (1980) was able to cite only one case of non-biological oxidation of herbicides (amitrole) although several cases of organophosphate oxidation have been reported (Faust and Suffet, 1966; Getzin and Chapman, 1960; Spencer *et al.*, 1980). Oxidation of herbicides may also be a result of photochemical reactions as mentioned by Crosby (1983).

Herbicides were shown to undergo reactions other than hydrolysis with soil organic matter (Stevenson, 1972). Phenoxyalkanoic acids and phenyl-carbamic acids reacted with amino and CO groups, respectively, in soil organic matter.

Other reactions are possible for soil-applied herbicides. Young and Khan (1978) have shown that the herbicide glyphosate nitrosated to *N*-nitrosoglyphosate in soils in the presence of nitrate. Glyphosate is relatively persistent when applied to irrigation water, and under certain conditions nitrate can accumulate in soils so that the possibility for this reaction to occur is quite high. Furthermore, *N*-nitroso derivatives of some pesticides have been found to be either carcinogenic, mutagenic, or both (Elesparu *et al.*, 1974).

## D. Photodecomposition

Photodecomposition is one of the pathways of pesticide conversion in the environment and is induced mainly by ultraviolet (UV) sunlight radiation. In the UV region the energy adsorbed by organic molecules causes excitation of electrons and can induce such transformations as the breakage and/or formation of chemical bonds, fluorescence, etc. It has been proven that compounds of several classes of herbicides are decomposed by UV light. Energy absorption by herbicides is dependent upon their chemical structure; it is well known that compounds having an aromatic ring exhibit intense absorption in the UV region. Several herbicides absorb energy strongly at wavelengths between 220 and 380 nm, which insures the basic requirements for the sunlight photolysis process (Table 5). The compre-

**Table 5.** Adsorption Maxima in Water of Selected Herbicides[a]

| Herbicide | Absorption maximum (nm) |
|---|---|
| Simazine | 220 |
| 2,4-D | 220,230,283 |
| 2,4,5-T | 220,289 |
| IPC | 234 |
| Monuron | 244 |
| Propanil | 248 |
| Dicryl | 258 |
| Amiben | 297 |
| DNBP | 375 |
| Trifluralin | 376 |

[a]After Bailey and White (1965).

hensive review on herbicide photodecomposition published by Crosby and Li (1969) gives a good basis for the understanding of the mechanism and dimension of this process.

Under the influence of light, herbicides decompose in aqueous solution and in the air where short wavelength UV reaches herbicide molecules directly. Herbicides reaching the soil are much less exposed to the UV light, so that the photochemical effect decreases. The experiments carried out in one of our laboratories (B.Y.) showed that photochemical decomposition on adsorbed particles is practically nil. The investigations on photochemistry at surfaces and interfaces with application to soils are only at the beginning; however, there are some indications that ultraviolet radiation could alter the behavior of adsorbed substances (Spencer *et al.*, 1980)

Fluchloralin was found to undergo photodealkylation on soil thin-layers (Nilles and Zabik, 1974) as was profluralin (Guth, 1980). In slightly alkaline soils only oxidation of the trifluoromethyl group and formation of non-extractable residues was observed for profluralin. Bentazone photolysis on soil surfaces occurs but no clear pathways could be established (Nilles and Zabik, 1975), although oxidative dimerization and loss of $SO_2$ was observed. Guth (1980) reported on the photochemical degradation of simazine, propazine, atrazine, and terbuthylazine on soil and found that none of the *N*-dealkylated products typical of *s*-triazine photolysis in water were produced. Since his control samples behaved similarly, it is questionable if photochemical decomposition occurred.

An interesting case of surface-induced photolysis of paraquat was described by Slade (1966). Paraquat solutions do not degrade in sunlight because paraquat's UV adsorption maximum, 257 nm, is very narrow and lies outside the range of UV radiation in sunlight. When paraquat is adsorbed on soil surfaces, its adsorption maximum shifts to 275 nm and becomes broader so as to include the lower end of the sunlight spectrum, thus enabling photochemical decomposition to occur.

In irrigated agriculture, herbicides may undergo photochemical decomposition under specific conditions such as in conveyance channels or in water in flooded paddy fields or sprinkler irrigation. In a flooded soil, as in a conveying channel, the applied herbicides are partitioned between water and the solid phase consisting of suspended particles. The water phase contains organic solutes that may act as sensitizers inducing the photochemical decomposition of herbicides (Zepp *et al.*, 1977). Consequently, in flooded systems the photodecomposition of herbicides may proceed either by direct UV adsorption or by formation of singlet-oxygen as a result of the presence of natural sensitizers or by a heterogeneous photocatalysis at the interfaces.

## E. Volatilization

Volatilization is defined as the loss of chemicals from surfaces in the vapor phase, e.g., vaporization followed by movement into the atmosphere. Volatilization of pesticides is an important pathway for their loss from treated agricultural lands. Reviews on volatilization include those by Hamaker (1972b), Spencer *et al.* (1973, 1982), Wheatley (1973), Guenzi and Beard (1974), Plimmer (1976), Taylor (1978), and Hance (1980b). Factors controlling pesticide volatilization have been extensively studied under controlled conditions in the laboratory (Spencer and Cliath, 1973; Jury *et al.*, 1980). In recent years, several studies of actual volatilization rates of pesticides under field conditions using the aerodynamic approach have provided an assessment of the rate of input to the air under typical conditions of use. In the aerodynamic method, pesticide volatilization losses are calculated from pesticide concentration, wind speed, and temperature gradients in the atmosphere that are obtained simultaneously over the field surface (Parmele *et al.*, 1972).

### 1. The Volatilization Process

Each chemical has a characteristic saturation vapor pressure or vapor density that varies with temperature. Potential volatility of a chemical is related to its inherent vapor pressure, but actual vaporization rates depend on the environmental conditions and all other factors that control behavior of the chemical at the solid–air–water interface. For surface deposits, the actual rates of loss, or the proportionality constant relating vapor pressure to volatilization rates, are dependent upon external conditions that affect movement away from the evaporating surface, such as turbulence, surface roughness, windspeed, etc. The rate at which a pesticide moves away from the surface is diffusion controlled. Close to the evaporating surface there is relatively no movement of air, and the vaporized substance is transported from the surface through this stagnant air layer only by molecular diffusion. Since the thickness of the stagnant boundary layer depends on air flow rate and turbulence, vapor loss is influenced strongly by the type of soil cover and the atmospheric conditions, e.g., wind in the vicinity of the soil surface. In general, under a given set of conditions, as air exchange rate of air tubulence increases, volatilization rate increases.

Volatilization of herbicides from soil is much more complicated and difficult to predict because of the many parameters affecting their adsorption, movement, and persistence. Soil-incorporated pesticides volatilize at a rate dependent not only on their equilibrium distribution between the air, water, and soil matrix as related to vapor pressure, solubility, and adsorption coefficients, but also on their rate of movement to the soil surface. Volatilization from soil involves desorption of the pesticides from the soil, movement to the soil surface, and vaporization into the atmosphere. Volatilization rates of soil-incorporated pesticides are dependent upon the vapor pressure of the pesticide in soil and the rate of movement of the pesticide to the soil surface. Vapor pressures of pesticides are greatly decreased by their interactions with the soil, mainly resulting from adsorption. Spencer *et al.* (1969) reported that the degree of reduction in vapor pressure in soil caused by adsorption is dependent mainly upon soil water content, the nature of the pesticide and its concentration, and soil properties, particularly soil organic-matter content. The concentration of the desorbed pesticide in the soil water dictates the vapor density of the pesticide in the soil air in accordance with Henry's law. Hence, the soil water adsorption coefficients can be used to calculate vapor densities in the soil atmosphere.

Soil water contents effects are especially important in adsorption of relatively non-polar pesticides by soil. Measurements of vapor pressure of pesticides in soil at various water contents conclusively demonstrated that the greater vaporization from wet than from dry soils is due mainly to an increased vapor pressure resulting from displacement of the chemical from the soil surfaces by water. This is illustrated in Figure 7, which shows the effect of soil water contents on vapor density of trifluralin in Gila silt loam, a coarse-loamy, mixed (calcareous), thermic Typic Torrifluvent (Spencer and Cliath, 1974). At the highest soil water content of 19%, the trifluralin vapor density or potential volatility was 3000 to 5000 times greater than when the soil was air dry. This great reduction in vapor density with soil drying accounts for the observed field measurements of trifluralin volatilization rates under extreme drying conditions reported by Harper *et al.* (1976) and Glotfelty *et al.* (1984).

## 2. Published Results

Volatilization rates from plant or moist soil surfaces can be very large, with losses approaching 90% within 3 days for more volatile pesticides (Taylor,

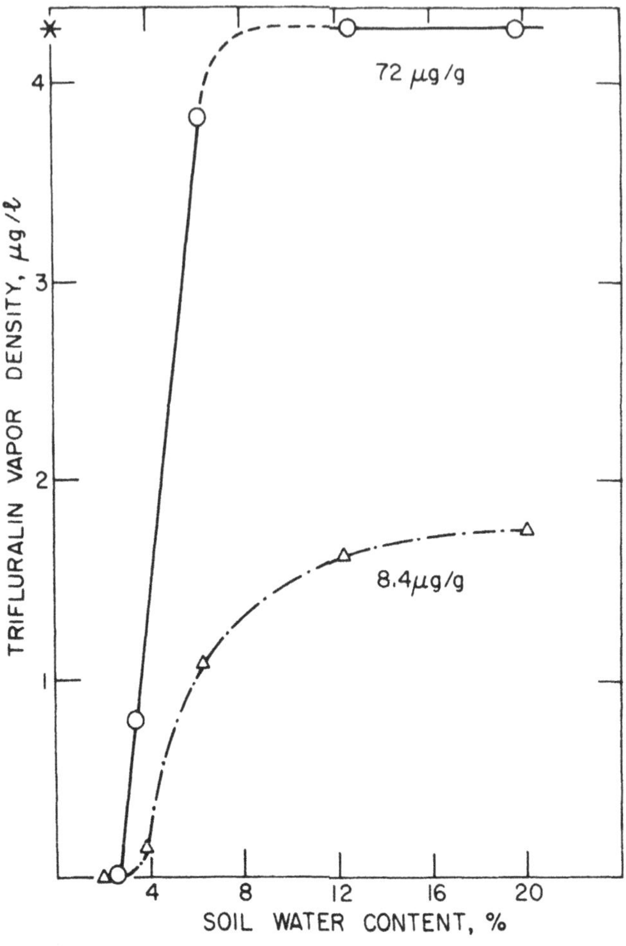


**Figure 7.** Effect of soil water content on vapor density of trifluralin in Gila silt loam at 8.4 and 72 μg/g trifluralin and 30°C. Reprinted with permission from Spencer and Cliath, *J. Agr. Food Chem.* 22: 987–991. Copyright 1974, American Chemical Society.

1978). Volatilization losses from dry soil or from soil-incorporated chemicals are much less. Measured volatilization rates three orders of magnitude less than those for exposed residues of the same pesticide have been reported.

Several herbicides were included in the field measurements of volatilization. For example, Harper *et al.* (1976) and White *et al.* (1977) measured trifluralin volatilization losses from a soybean field after shallow incorporation. Seasonal trifluralin volatilization losses were estimated to be 22.4% of that applied. They reported that trifluralin losses decreased to very low levels during daytime when the surface soil water content was low, even though turbulence, soil temperature, and evaporative demand were high. They indicated that when the soil surface was not wet, trifluralin losses appeared to be controlled by surface soil water content and its effect on pesticide adsorption to the soil particles. Turner *et al.* (1978) compared volatilization of microencapsulated and emulsion formulations of chlorpropham following application to a bare silt loam soil without incorporation. Over a 50-day period following application, the emulsified residues volatilized about five times faster than the encapsulated chlorporpham. Glotfelty *et al.* (1984) measured volatilization of trifluralin and dacthal herbicides following spraying on moist, bare soil surfaces of two soils. Initially rapid volatilization rates from the moist silt loam soil were proportional to the saturation vapor densities of the pure compounds, with half of the trifluralin disappearing in less than 3 days. The maximum measured trifluralin volatilization rate was equivalent to 4.7 kg $ha^{-1}day^{-1}$. Volatilization of dacthal was much less, only 2% during the first 34 hrs following application. Loss rates from a very sandy soil were much lower, probably because of a dryer soil surface.

Grover *et al.* (in press) reported that volatilization loss of 2,4-D in 3 days after application of the isooctyl ester to a wheat field was equivalent to 20% of the amount applied. Cliath *et al.* (1980), in the only direct measurement of herbicide volatilization from an irrigated field, measured volatilization losses of EPTC from an alfalfa field following its application at 3 kg $ha^{-1}$ in irrigation water with a flood irrigation system. They reported volatilization losses of 74% of the applied EPTC from irrigation water and moist soil within 52 hrs following its application. Their data indicate that using surface irrigation water to apply highly volatile herbicides is an extremely inefficient method of application considering the high losses resulting from volatilization.

### 3. Estimating Volatilization Rates from Soils

Volatilization of herbicides from soils can be estimated from a consideration of the physical and chemical factors controlling their concentration at the soil surface. With soil-incorporated herbicides, the initial volatilization rate will be a function of the vapor pressure of the chemical at the surface as modified by adsorptive interactions with the soil. The small fraction of the exposed material that remains on the soil after mixing is readily lost. Volatilization then becomes dependent upon the rate of the movement of the pesticide to the soil surface by diffusion and convection in evaporating water. Usually both mechanisms, diffusion and convection, work together in the field where water and the pesticide vaporize at the same time. Movement of the pesticide to the surface by bulk flow or convection in the soil water is the dominant factor in controlling volatilization of pesticides incorporated in moist soil. Most models developed for estimating volatilization rates are based upon equations describing the rate of movement of the chemicals to the surface by diffusion and/or by convection and away from the surface through the air boundary layer above the surface by diffusion.

Additionally, the proportion of a pesticide in soil that will be lost by volatilization depends upon the resistance of the chemical to degradation.

In a series of papers Jury *et al.* (1983a, 1983b, 1984 a, b) described and applied a model for assessing relative volatility, mobility, and persistence of pesticides and other trace organics in the soil. The model assumes that the soil-surface boundary consists of a stagnant boundary layer connecting the soil and air through which pesticides and water vapor must move to reach the atmosphere. It further assumes that the gas and liquid concentrations are related by Henry's law, that the adsorption isotherms are linear, and that degradation occurs by a first-order rate process. The model (Jury *et al.*, 1983a) is intended to classify and screen organic chemicals for their environmental behavior based on their physical and chemical properties such as vapor pressure, solubility, or Henry's constant, organic carbon partition coefficient, and degradation rate.

Jury *et al.* (1984b) applied the screening model to a set of 20 pesticides and 15 other trace organic chemicals for which benchmark properties were obtained from the literature or calculated. The model indicated that volatilization behavior of a chemical was controlled mainly by the ratio of its solution to vapor concentration or Henry's constant, which determined the extent to which the air boundary layer restricted the volatilization from soil. The extent to which this boundary layer limits volatilization can be used as a criterion for classifying pesticides into general categories, based upon whether control of volatilization is within the soil or within the boundary layer. When a boundary layer is present it will act to restrict volatilization only if the maximum flux through the boundary layer is small compared to the rate at which chemicals move to the soil surface. This independence or dependence on boundary layer properties has implications for volatilization of soil-incorporated pesticides. Assuming category I chemicals are not affected by the boundary layer and category III chmeicals are dominated by the boundary layer properties, a clear-cut distinction in volatilization between category I and III chemicals is predicted. For category I chemicals, such as trifluralin, the volatilization flux will decrease with time even in the presence of vaporizing water. Category III chemicals with a low Henry's constant, such as atrazine, move to the surface in evaporating water faster than they can volatilize into the atmosphere; consequently, their concentration increases at the soil surface under evaporative conditions and volatilization rate increases with time. The screening models developed by Jury *et al.* (1983a) should provide a useful mechanism for determining the relative importance of volatilization and other pathways for loss of applied herbicides in soils. The model should also be useful for assessing the relative loss by volatilization of new herbicides based on their physico-chemical properties.

Since the volatility of herbicides in soil is so greatly affected by the soil moisture status, irrigation management can be an important factor in controlling loss of herbicides by volatilization. Proper irrigation management can increase efficacy of herbicides and also hasten their loss to prevent residual carryover to the next crop. If herbicides are applied to moist soils following irrigation, excessive volatilization may occur before they can be incorporated. On the other hand, irrigations can be used to move applied herbicides into the soil to prevent loss by volatilization. Irrigations near the end of the season can be used to enhance volatilization of excess herbicide to prevent carryover to the next crop.

## IV. Herbicide Transport

The transport of herbicides in soil is governed by both soil and chemical properties. As previously stated (section IID and IIE), irrigation technology, fluctuations of soil moisture content, and temperature are additional factors that may affect the transport of herbicides into the soil or within the unsaturated zone between the soil surface and the ground-water table.

Two patterns characterize solute movement in the soil. In the first one, the dominant transport process is liquid diffusion typical of herbicides with low water solubility and high adsorption capacity on soil surfaces. In the second one, characterizing movement of herbicides soluble in water, mass flow in the soil water phase is the predominant process. Hydrodynamic dispersion—including both diffusion and mass flow—defines solute spreading in soil and other porous media owing to water velocity variations. At low average water fluxes in uniform soil, this process is relatively unimportant but becomes dominant over diffusion at high water fluxes or in structured soils where substantial variation in water velocities exist. Herbicides, being adsorbed on the soil solid phase, do not move freely with water through the soil. For example, Jury (1983) showed that the travel time for an adsorbed chemical, $t_A$, is related to the travel time for a non-adsorbed or mobile chemical, $t_M$, by

$$t_A = (1 + \rho_b R/\theta)\, t_M \equiv K t_M, \qquad [3]$$

where $\rho_b$ is the soil dry bulk density, $K$ is the partition or distribution coefficient of the interacting chemical, $\theta$ is volumetric water content, and $R$ is a generalized sink term modified for specific reactions (mass of solute/volume/time).

Under field conditions, the movement of herbicides often does not follow the anticipated general pattern. Soils with a high content of clay when subjected to wetting and drying cycles may shrink and crack. Then when a herbicide is applied, followed by an irrigation, the herbicide will partially leach into the soil through cracks and large pores. During transport through the large pores or cracks only a portion of the solid phase—the external surface—comes in contact with the solute and the amount retained by the soil is relatively small.

Herbicides may be transported also when adsorbed on soil particles. The main pattern of such a transport is in runoff, when the adsorbed chemicals are laterally transported over the soil surface. Its magnitude is defined by the intensity of the falling water, the soil infiltration capacity, and the slope of the surface. Also vertical transport of herbicides adsorbed on suspended particles may occur through cracks and large pores in soils on which herbicides have been applied.

Modeling of solute transport into the soil has been a central interest of the scientific community in the last decade, and part of this interest concerns application of models in predicting herbicide movement into soil. Rao and Jessup (1982) point out, however, that "the ability to formulate and numerically solve complex simulation models apparently exceeds our ability to verify these models using the limited experimental data base presently available and transferring the existing parameters to a field scale soil heterogeneity."

## A. Diffusion

Herbicides can diffuse through soil in both vapor and liquid phases. Diffusion is the process whereby material moves from a higher to a lower concentration as a result of random molecular motion. The rate of transfer of material is proportional to the concentration gradient normal to the direction of movement. The basic mathematics of the process are largely described in Crank (1975), and its application to biocide behavior may be found in Hamaker (1972b), Letey and Farmer (1974), Hartley and Graham-Bryce (1980), Calvet (1984), and Jury *et al.* (1984b). Since diffusion in the liquid phase is the dominant one involved in the transport of herbicides under irrigation, only this aspect will be discussed here.

Letey and Farmer (1974) described conditions in the soil system that may affect solute diffusion in soil as follows: (1) The diffusion coefficient cannot be assumed to be independent of concentration; (2) diffusion is confined to certain segments of the system (a molecule will not diffuse through a solid soil particle); (3) sorption of the diffusing substance by soil particles often occurs; and (4) the diffusion coefficient is dependent upon temperature and a number of soil properties such as mineral composition, bulk density, and water content. If it is assumed that herbicide movement by volatilization is negligible, no degradation occurs during the diffusion process, and the adsorption is linear and single valued, then the Olsen and Kemper (1968) relationship

$$D = \frac{D_0 f \theta}{(\rho K + \theta)} \qquad [4]$$

expresses the diffusion coefficient ($D$).

The variables $K$ (adsorption from aqueous solution), $\rho$ (dry soil bulk density), $f$ (impedance factor that takes into account the tortuous pathway followed by the solute through the soil pores), and $D_0$ (diffusion coefficient of the chemical in water) are empirically determined.

Gerstl *et al.* (1979b) found that a mobile fraction of the adsorbed phase of an organic molecule may also contribute to diffusion. The adsorbed molecules may be divided into an immobile fraction and a mobile fraction contributing to diffusion and having the same apparent mobility as molecules in solution. If a fraction α of the adsorbed molecules is mobile, than equation [4] becomes

$$D = \frac{D_0 f(\theta + \alpha\rho K)}{(\rho K + \theta)}. \quad [5]$$

Since diffusion of organic molecules in soil remains at low values even over a large range of moisture content, biological degradation may occur during the diffusion process. Gerstl *et al.* (1979a) proposed a model to include microbial activity in the diffusion system that considers rate of decomposition at any distance and time dependent on the local concentration of the chemical and on microbial activity. The model predictions satisfactorily fit the experimental results.

Several researchers have measured diffusion of herbicides through soil at various soil water contents. Lavy (1970) measured the diffusion coefficient of three *s*-triazines in six soils at two water contents. The average diffusion coefficients reported were $0.26 \times 10^{-8}$ and $1.83 \times 10^{-8}$ $cm^2\ sec^{-1}$ at one-half field capacity and field capacity, respectively. Scott and Phillips (1972) measured the diffusion of several herbicides as affected by water content. The diffusion coefficients for prometone, atrazine, simazine, prometryn, diphenamid, and chlorpropham increased with increases in soil water content (Table 6).

The effect of adsorption on diffusion of 2,4-D was measured by Lindstrom *et al.* (1967) and of propazine and prometryne by Walker and Crawford (1970). In both cases, an inverse relationship between measured coefficients and adsorption was observed. The temperature effect was measured by Lavy (1970). He reported a decrease in the diffusion coefficients of the *s*-triazine herbicides (atrazine, propazine, and simazine) as soil temperatures decreased from 50 to 25°C.

Since diffusion of herbicides through soils is governed by the soil water content, it is obvious that this process may be controlled in irrigated agriculture where fluctuations in soil moisture content area are part of the soil-crop management system.

### B. Mass Flow

As water moves through the soil, dissolved herbicides are carried along in the convective stream. The mass transport of water and solute in soils takes place through a complex three-dimensional network of pores characterized by an infinite number of forms, sizes, and shapes. During their flow through

**Table 6.** Diffusion Coefficient ($D$) of Some Herbicides at Various Soil Water Content ($\theta$)[a]

| Herbicide | $D \times 10^{-7}$ cm$^2$ sec$^{-1}$ | $\theta$ |
|---|---|---|
| Prometon | 1.2 | 0.221 |
| | 3.65 | 0.325 |
| | 8.9 | 0.494 |
| Atrazine | 1.1 | 0.325 |
| | 4.25 | 0.494 |
| Simazine | 0.5 | 0.221 |
| | 1.25 | 0.325 |
| | 3.8 | 0.494 |
| Prometryn | 0.55 | 0.325 |
| | 4.15 | 0.404 |
| Chlorpropham | 0.8 | 0.221 |
| | 1.0 | 0.325 |
| | 7.15 | 0.494 |
| Diphenamid | 2.1 | 0.325 |
| | 2.55 | 0.390 |
| | 3.1 | 0.494 |

[a]As measured by Scott and Phillips (1972).

soil pores, the herbicides are subject to a series of interactions like adsorption, degradation, volatilization, and plant uptake. Under conditions of unsteady or transient water flow, the contact time between the solute and the soil solid phase often is insufficient to achieve an equilibrium. This condition is characteristic of the simultaneous flow of water and solute during the infiltration/redistribution process as a result of irrigation. Rao and Jessup (1983) describe the convective-diffusive, dispersive transport of pesticides during transient water flow by the equation

$$\frac{\delta}{\delta t}(\theta C + \rho S) = \frac{\delta}{\delta x}\left[D_h \theta \frac{\delta C}{\delta x}\right] - \frac{\delta}{\delta x}(v\theta C) - \sum_{i=1}^{n} Q_i, \qquad [6]$$

where $\rho$ is soil bulk density, $D_h$ is hydrodynamic dispersion coefficient, $v$ is average pore-water velocity, $Q_i$ are various sink terms to account for pesticide losses (degradation and plant uptake), $S$ and $C$, respectively, adsorbed phase and solution phase concentrations, $\theta$ is volumetric soil water content, and $x$ is soil depth.

The reader is referred to reviews published by Bailey and White (1970), Letey and Farmer (1974), Leistra (1978, 1980), Jury *et al.* (1983a), Rao and Jessup (1982, 1983) for a comprehensive picture and the detailed mathematics of convective transport of biocides through soils.

The movement of some herbicides through soil as a mass flow process was measured both under laboratory (column) and field conditions. Hornsby and Davidson (1973) measured the distribution of fluometuron in a soil column for various displacement periods and water flow velocities. van Genuchten *et al.* (1974) reported results of studies on picloram mobility in soil columns. Using miscible displacement techniques, Davidson and Chang (1972) measured the effect of water flow rate, soil bulk density, and aggregate size on transport of picloram and reported that a tenfold decrease in water velocity caused a significant decrease in picloram mobility.

When high liquid fluxes are used, the peak of herbicide distribution moves faster than expected from equilibrium adsorption. This fact was observed with prometryn and fluometuron (Abernathy and Davidson, 1971; Davidson and McDougal, 1973). Wu *et al.* (1975) observed that a small fraction of applied napropamide moved very fast into the soil with water percolated at a high rate under saturated conditions. The same phenomena was observed in an experiment carried out with napropamide by White *et al.*[1] on structured, undisturbed Oxford soil columns. Those findings confirm the results of Davidson and Chang (1972) who applied herbicides in solution to soil columns consisting of either small or large aggregates and found that at high liquid fluxes the non-equilibrium effects were more marked with larger than with the smaller aggregates. This was explained by the longer diffusion time required for the solute in the liquid phase to reach equilibrium with the internal portion of the larger aggregates. If start of column leaching is delayed following herbicide incorporation, the herbicide has sufficient time to penetrate into small pores and to be adsorbed onto solid surfaces (Leistra, 1980). Then, when water is applied at a high rate, the adsorbed substance will only slowly be released into the flow path so the main part of it will move substantially slower than it would under equilibrium conditions.

Type of soil, flow rate, and the experimental techniques lead to some inconsistency regarding the relative mobility of herbicides. Weber and Whitacre (1982) selected four herbicides—atrazine, diuron, prometon, and bromacil—to follow their relative mobilities as described by various authors. The results were as follows: atrazine > diuron (Harris, 1966; Helling, 1971) atrazine > prometon > atrazine (Helling, 1971); bromacil >> diuron (Rhodes *et al.*, 1970)

A multifactorial column experiment was recently reported by Weber and Whitacre (1982) to determine (1) the relative mobility of atrazine, bromacil, buthidazole, diuron, and prometon in Lakeland sand under unsaturated

[1]Personal communication, 1984.

flow conditions, (2) the relative mobilities of buthidazole, tebuthiuron, and CN-10-3510 in Lakeland sand (a thermic, coated Typic Quartzipsamment) under saturated flow conditions, (3) the effects of three different types of leaching methods (unsaturated flow, saturated-unsaturated flow, and saturated flow) on mobility of tebuthiuron in Lakeland sand, (4) the mobility of tebuthiuron applied to Lakeland sand and maintained in a moist state or oven-dried and leached under unsaturated-flow conditions, and (5) the mobility of tebuthiuron in three soils under unsaturated-flow conditions. The above researchers found that under unsaturated-flow conditions, bromacil was considerably more mobile than buthidazole, and because of their high water solubilities, both herbicides were much more mobile than atrazine, prometon, or diuron. Under saturated-flow conditions, buthidazole was leached through Lakeland loamy sand in slightly greater amounts than tebuthiuron or CN-10-3510. Distribution of the three herbicides in the leached soil was similar and relatively uniform. In Lakeland loamy sand, 30 times as much tebuthiuron was leached under saturated-flow conditions as under unsaturated-flow conditions. Intermittent saturated–unsaturated-flow conditions resulted in four times as much leaching of tebuthiuron as unsaturated flow alone. Only one-tenth as much tebuthiuron leached under intermittent saturated–unsaturated-flow conditions as under saturated-flow conditions.

Measured transport of herbicides in the field often deviates from the expected pattern. Redistribution of the applied water resulting from evapotranspiration, presence of cracks and voids, or heterogeneity of the soil constituents may explain this deviation. Schiavon and Jacquin (1973a, 1973b) reports the distribution of atrazine following application to a clay and clay loam soil at 1.25 kg/ha. In those soils characterized by swelling and shrinkage, low amounts of atrazine were transported fairly rapidly to a depth of a few decimeters with the highest concentrations remaining in the upper 6 cm of soil. Winter rain led to the leaching into the drain of about 2% of the applied atrazine. This can be explained by shrinkage and cracking of the soils during the dry period favoring a preferential pass flow of water and solute. Bovey *et al.* (1975) reported the results of a study where 2,4,5-T and picloram were applied on slowly permeable montmorillonitic clays with pronounced shrink-swell characteristics. Essentially no 2,4,5-T or picloram was found in seep (subsurface) water from an area treated five times with the above herbicides during a 3-yr period. Small amounts (1–2 ppb) of picloram were detected in the drainage water at the 130-cm depth with the amount increasing 3 to 4 ppb when supplemental irrigation was applied. In this situation, the movement of both herbicides was retarded by adsorption.

Data of Clay and Scott (1973) on simazine, Baumaister and Hurle (1975) on atrazine, and Leistra *et al.* (1975) and Marriage *et al.* (1977) on the uracil herbicides show the effect of adsorption on retarding vertical transport of herbicides under various soils and environmental conditions.

### C. Movement in Runoff Water

Runoff water is an important means of herbicide transport from one soil location to another or from the soil to surface water bodies. Since runoff can be induced either by rains or irrigation, the process is included in the present paper.

What is the difference between the transport of herbicides in runoff water compared with soil water? Usually in soil water herbicides are transported as solutes only and the governing factor in their transport is their water solubility and adsorptive capacity; whereas in runoff water herbicides are transported both as solutes and adsorbed on sediment particles, and in this way highly insoluble and adsorbed herbicides may move also.

In addition to chemical and soil properties, the amount of pesticides found in runoff water is affected by the intensity of falling water and slope of the land. The formulation of the herbicide and its method of application may also affect its transport by runoff. Partitioning between the solid and liquid phases will define the ratio of the amount of herbicide transported as solute and the amount adsorbed on suspended particles.

Review articles were published in the last decade by Merkle and Bovey (1975), Wauchope (1978), and Hartley and Graham-Bryce (1980), and the reader is referred to the above materials for a detailed discussion of pesticide transport in runoff water. In this paper we selected to present only a few examples that illustrate the partitioning of some herbicides between water and sediment in the runoff water.

The distinguishing characteristic of herbicide transport by runoff is the simultaneous movement of the chemical as solute and on suspended particles. Hall *et al.* (1972) evaluated the presence of atrazine in runoff water in solution and on eroded soil sediments. The pattern of loss of atrazine in sediment throughout the growing season was similar to that found for runoff water, i.e., the levels of atrazine showed a steady decline with each succeeding runoff event. The level of atrazine in sediment was greater than in solution, as would be expected since atrazine molecules are highly adsorbed on soil colloidal materials. Leonard *et al.* (1979) studied atrazine, cyanazine, diphenamid, paraquat, propazine, and trifluralin in watershed runoff from rainfall. He found that paraquat was transported only on adsorbed sediments and not in solution in the runoff water. Diphenamid and atrazine were transported mainly as solutes. In a recent study on irrigation-induced runoff of herbicides, Spencer *et al.* (in press) determined the partitioning of the three chemicals between sediment and water as follow: DCPA 4.6% in water, 95.4% in sediment; prometryn 100% in water; trifluralin 34.4% in water, 65.6% in sediment.

The chemical formulation is an additional factor affecting the transport of herbicides in runoff. Wauchope (1978) showed that, for both seasonal and long-term studies (1) wettable powders consistently produce the highest runoff losses, up to 5% with amounts of the application, (2) water-insoluble pesticides—usually applied as emulsions—show losses of 1% or less, and (3) water-soluble pesticides—usually applied as aqueous solutions—and soil-incorporated pesticides show losses of 0.5% or less.

## D. Modeling

The need for predicting the simultaneous transport of water and herbicides in soil for better crop protection and management and for the prevention of unsaturated zone and surface water pollution has stimulated the development of a series of mathematical models. In general, the existing models for the transport of non-interacting solutes were adapted to herbicide properties—adsorption, degradation, volatilization, and runoff—and specific models were produced that, in general, provide a partial agreement with experimental data. For a comprehensive discussion of the topic, the reader is referred to the papers of Leistra (1973, 1977, 1978, 1980); Davidson and McDougal (1973), Hornsby and Davidson (1973), van Genuchten *et al.* (1974, 1977), Haque and Freed (1975), Rao *et al.* (1979), Rao and Davidson (1980), Gerstl *et al.* (1979a, 1979b), and Jury *et al.* (1983a).

A critical review of the development and verification of existing simulation models for describing pesticide dynamics in soil was prepared recently by Rao and Jessup (1982). It is not our intent to repeat or extend the discussion of modeling transport processes. However, part of the existing mathematical models being developed for equilibrium condition are inadequate for field predictions. Even though batch experiments suggest a rapid equilibrium of adsorption, in reality during transient conditions the reactions are only partially completed. The models proposed by van Genuchten and Wierenga (1976) and by Leistra (1977) consider that the rate of pesticide adsorption-desorption in "sites residing within immobile water" is controlled by diffusive mass transfer across the liquid layer. This results in an incomplete equilibrium during solute flow through soils. This situation is compounded by the heterogeneity of field scale where substantial lateral variations in vertical water velocity can be found. Jury (1983), in summarizing the current approaches in chemical transport modeling, stated one unresolved problem as follows: "convective models such as the transfer function model which take velocity variations into account are able to describe the spatial variability of chemical movement near surface but it is unknown to what depth the correlation inherent in the assumption of the transfer function model will still hold."

Attention is presently being given to developing a model for transport of herbicides by runoff under irrigation (Spencer, in press). The pesticide submodel used in CREAMS (Knisel *et al.*, 1980) for runoff from rainfall areas is considered a starting point in developing a submodel routine for surface-irrigated fields since many routes of dissipation from the soil are similar in irrigated and rainfall areas.

Herbicides are among the most interacting chemicals with the soil medium, being subject to adsorption–desorption, degradation, conversion, volatilization, and runoff. Consequently, it is understandable that a large number of parameters are required in a valid model able to simulate their transport into or over soil.

## V. Irrigation Practices and Herbicide Behavior

As previously stated, besides the properties of the soil and the chemicals, the fate of herbicides is governed by the application method and irrigation practices. Herbicides may be directly applied to the soil surface and then irrigated, applied into the soil together with the irrigation water, or sprayed on flooding water.

Examples of herbicide fate in the above three irrigation systems will be presented and discussed.

### A. Fate of Herbicides in a Flooded Soil

Application of herbicides to flooding water is practiced in rice culture and in open channels conveying water to irrigated fields. Herbicides in rice culture were recently the subject of a special symposium in the framework of the Fifth International Congress of Pesticide Chemistry where comprehensive review papers were presented on the above topic (Crosby, 1983; Chen, 1983; Kuwatsuka, 1983). The fate of herbicides in open channels was discussed in various research papers (Demint *et al.*, 1970; Comes *et al.*, 1975; Hiltibran and Turgeon, 1977), and the findings were similar to those characterizing paddy soils.

Crosby (1983) considers the rice field as a chemical reactor containing aqueous solution, reactive nucleophiles such as hydroxides characterized by a pH between 8 and 10, ammonia, amines, sulfides, dissolved oxygen and other oxidants, organic and inorganic reducing agents, and biochemical compounds resulting from intense microbiological activity. The energy is provided by heat and light. Herbicides released into this reactor are subjected to dissolution, adsorption, leaching, runoff, volatilization, and biological, chemical, and photochemical conversion.

Adsorption is inversely related to water solubility and so most rice herbicides are not considered to be highly adsorbed, although some may become chemically bound on the soil surface. Volatilization may be one of the main dissipation pathways in a flooded soil. Since the rate of volatilization in water is affected by solubility, vapor pressure, and water depth, the vapor loss from a shallow and warm flooded field can be highly significant (McKay and Leinonen, 1975). Leaching of herbicides in flooded paddies should be greater than in upland fields because in submerged soils they will be continuously leached with the percolating water. Runoff of paddy herbicides or of applied chemicals in water channels is one of the main pathways of dissipation characterizing the behavior in a flooded ecosystem.

Photochemical and biochemical degradation govern the fate of herbicides in a flooded environment. The high pH of the water induces hydrolysis of carboxylic esters, but other chemical reactions are relatively slow. Some herbicides can adsorb UV radiation. These light-energized molecules can undergo reactions resulting primarily in oxidation (oxygenation) and photonucleophilic replacement of substituents by hydroxyl derived from water (Crosby, 1983). The presence of natural sensitizers such as humic and fulvic acids in flooding water may also induce photochemical degradation of herbicides that do not adsorb UV radiation.

Microbial degradation of herbicides in water is less important than other pathways of loss in the flooded environment. However, where herbicides are applied pre-emergence to the soil, microbial degradation may become important after flooding (Matsunaka and Kawatsuka, 1975). A summary of the existing knowledge on the residual periods of herbicides in a flooded ecosystem was presented recently by Kuwatsuka (1983). The half-life of paddy herbicides established in laboratory experiments and in fields is presented in Table 7. Despite the fact that the laboratory experiments carried out under controlled conditions show in general a greater persistence, a good agreement is observed with their fate in the paddy field.

Two compounds—benthiocarb and molinate—largely used in paddy soils in the U. S. and Japan were selected to illustrate the behavior of herbicides under flooded ecosystem. Benthiocarb or thiobencarb, which is a moderately adsorbed herbicide, was still retained mostly in the surface 2-cm layer of paddy fields after its application (Ishikawa *et al.*, 1976). They report that the residual amount of benthiocarb in soil in flooding water of paddy fields decreased rapidly one week after application. Climatic condition affects its persistence; for example, Kuwatsuka (1983) presenting an experiment done by Yamada *et al.* (1979) showed that benthiocarb residues in soils after harvest were around 20 to 50% in one cold area of Japan and 3 to 4% in a tropical area.

Crosby (1983) and Kuwatsuka (1983) emphasize that both photochemical and microbial degradation of benthiocarb occurs in paddy soils. Studies on the degradation of benthiocarb in soil carried out by Ishikawa *et al.* (1976) and Nakamura *et al.* (1977) show that phenyl—$^{14}C$—benthiocarb was readily degraded in upland soils under aerobic conditions, less readily in flooded soils under anaerobic-aerobic conditions, and very slowly in flooded soils under anaerobic conditions. Very little degradation of benthiocarb was observed in sterile soils. Ishikawa *et al.* (1976, 1980) and Nakamura *et al.* (1977) identified as degradation products of benthiocarb,

**Table 7.** Half-Life (Days) of Some Paddy Herbicides in Flooded Soils[a]

| Herbicide | In laboratory experiments[b] | In paddy fields[c] |
|---|---|---|
| 2,4-D | 30–40 | 33, 48 |
| 2,4-D ethyl | <1 hr | |
| MCPA | 3, 4, 7, 7, 15, 20 | 7, - |
| MCPA ethyl | 7–14 | |
| Phenothiol | <1 | .5 |
| Dalapon | 3–5, 14–21 | 10–15, 20–31 |
| PCP | 5, 10–17, 12–70 (mean 30), 60 | 3–10, 6–7, 10–17 |
| Benthiocarb | 7–100 (mean 40) | 3–8, 7, 8, 11, 62 |
| Molinate | 7, 15, 18, 30 | <1, <5, <5, <5, <9, 11 |
| Swep | 2–9, 10–11, 7–14 | <10 |
| Propanil | <1–1, 1, 5 | <1–1 |
| Naproanilide | 5, 6 | 2, <4, 4 |
| Napropamide | 49, 56 | <20, 20 |
| Butachlor | 9–30 | 6, 8 |
| Credazine | 90–150 | 22, 45 |
| Trifluralin | 9, 10, 22, 45 | 10, 45 |
| Nitrofen | 3–35 (mean 11) | ~14 |
| Chlornitrofen | 7–35 (mean 15), 17, 35 | 7, 9, 12, 13, ~14 |
| Chlomethoxynil | 7–35 (mean 15), 30 | 7–8 |
| Bifenox | 4, 4 | 4, 4 |
| Oxadiazon | 75, 93–98 | |
| Bentazone | 5, 33, 45 | 15, 15 |
| Simetryn | <37, 63 | |
| Prometryne | 100, 120 | |
| Paraquat | >180 | >180 |
| Diquat | >150 | |

[a]Data collected from various sources; after Crosby, 1983.
[b]Each chemical was mixed with soil and incubated at 25 to 30°C in dark.
[c]Persistence in soils in paddy rice fields in Japan. Most herbicides were applied in May or June.

benthiocarb sulfoxide, 2-hydroxybenthiocarb, desthylberthiocarb, 4-chlorobenzyl alcohol, 4-chlorobenzoic acid, 4-chlorobenzoic acid, 4-chlorobenzyl sulfoxide, and dechlorinated benthiocarb. Strains of coryneform bacteria that decompose benthiocarb were isolated; but a lag period is required before the microbial degradation of benthiocarb begins.

The photodegradation of benthiocarb in a paddy field was relatively high and photolysis products included *p*-chlorobenzaldehyde, *p*-chlorobenzyl alcohol, *p*-chlorobenzoic acid, and hydroxylated benthiocarb (Ishikawa *et al.*, 1976). This photodegradation seems "surprising" since in aqueous solution the herbicide did not adsorb sunlight UV. The only explanation for this phenomenon is that the flooding water in a paddy field contains humic and fulvic acids or other organic compounds that may act as natural sensitizers inducing photochemical decomposition.

Losses of benthiocarb occur also as a result of volatilization, leaching, and runoff. The herbicide is quite volatile. At 30°C pure benthiocarb was lost from aqueous solution with a half-life of 2 days; at 25°C, $t_{1/2}$ increased to 8 days. The rate of volatilization decreased when a soil layer was placed at the bottom of the solution because of the rapid adsorption of the herbicide on the soil particles. Leaching of benthiocarb was minute and not deeper than 5 cm. However, the presence of benthiocarb was reported in creeks as a result of the runoff of the herbicide through irrigation, rain, or overflow (Kuwatsuka, 1983).

Molinate is a non-persistent herbicide extensively used in paddy fields in recent years both in the U.S. and Japan. In paddy soils the molinate half-life is very short (Table 7). In addition to microbial degradation, it dissipates rapidly by volatilization and photodegradation (Soderquist *et al.*, 1977; Deuel *et al.*, 1977). The loss of molinate by runoff is higher than benthiocarb.

Two patterns characterize the fate of molinate in paddy fields. First is the degradation of the chemical in soil. It was recently reported that the degradation rate is a function of soil type and aerobic conditions (Thomas and Holt, 1980). The following degradation products in soil were identified: sulfoxide and sulfone of molinate, azepin-ring-hydroxy and 4-oxo derivatives, *s*-carboxy methyl derivative, *s*-2-hydroxyethyl derivative, and ethane sulfonate. Molinate degradation in soil is microbiological, and a large number of strains of degrading microbes were identified (e.g., *Mycobacterium* sp., *Flavobacterium* sp., *Sheptomycarb* of actinomycetes and *Fusarium* sp. of fungi). The degradation pathways differed largely according to the strain (Imai and Kuwatsuka, 1984). It was established that in paddy soils there are many molinate-degrading microbes and a lag period is not necessary before the microbial-induced degrading process begins.

The second pattern of molinate dissipation is photodecomposition–volatilization. Crosby and co-workers (1978, 1983), studying the connection between volatilization and photodecomposition of molinate, reached the conclusion that molinate does not absorb UV radiation at sunlight wavelengths and is relatively stable in the dark. However, like benthiocarb, it is rapidly photooxidized by hydrogen peroxide, and a number of photolysis products were detected in field tests. Owing to the high volatilization of molinate (volatilization only results in a $t_{1/2}$ of 25 days when applied in late May at 28°C) neither light nor microbes have time to contribute much to chemical conversion. Once in the atmosphere molinate is degraded by ozone and other photooxidants without adsorbing UV radiation. Figure 8 shows molinate photodegradation in water (w) and air (a). Molinate application leads to small decreasing amounts in water and

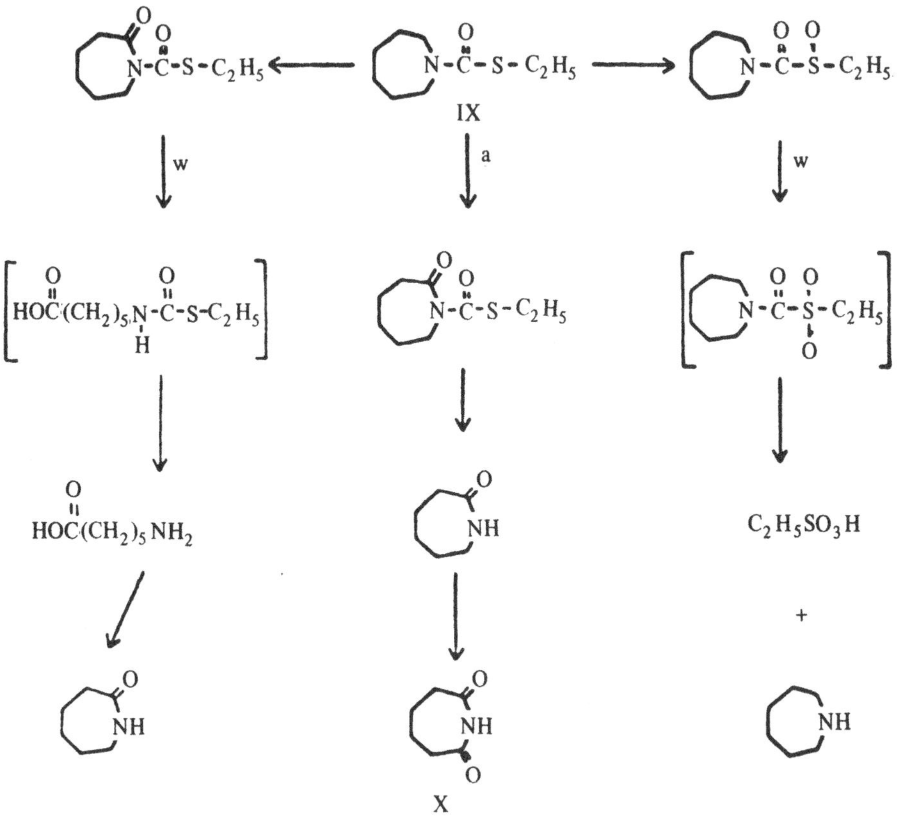


**Figure 8.** Photodegradation scheme of molinate in water (w) and air (a). From Crosby (1983).

increasing transport to the air phase because of volatilization. An estimate of the relative importance of each dissipation route made by Crosby (1983) and presented in Table 8 shows the significance of volatilization.

From the above examples, it may be observed that the fate of herbicides in flooded soils is a multifactorial process involving chemical and biological degradation, volatilization, photodecomposition in water and air, as well as vertical (leaching) and lateral (runoff) transport.

### B. Fate of Herbicides Applied Prior to Irrigation

The behavior of herbicide applied to a field prior to irrigation is defined not only by the soil's physico-chemical and hydraulic properties, but also by the irrigation regime (method, rate, amount, and frequency). The method of irrigation will affect the herbicide's spatial distribution, the rate and amount of irrigation will influence the extent of its dispersion, and the frequency of

**Table 8.** Fate of Molinate When Applied in Paddy Soils[a]

| Process | Estimated loss (%) |
|---|---|
| Soil adsorption and metabolism | <10 |
| Plant uptake and metabolism | < 5 |
| Aqueous microbial metabolism | < 1 |
| Hydrolysis | < 1 |
| Photolysis | 5–10 |
| Volatilization to atmosphere | 75–85 |

[a]After Crosby (1983).

irrigation will control the fluctuation of aerobic and anaerobic conditions in the field. Despite the fact that herbicides are applied in large amounts in irrigated agriculture, the information regarding their distribution as a result of irrigation technology is scarce. Parameters on herbicide vertical transport in soil have been obtained mainly in column studies, and data on ground-water pollution as a result of rainfall forms the bulk of the information dealing with herbicide leaching in the field (Leistra, 1980).

As was previously stated flooding, furrow, sprinkler, and drip irrigation are the main irrigation practices. Herbicide behavior in a flooded soil was already discussed, and in this part only cases of herbicide behavior in the field when applied prior to furrow, sprinkler, and drip irrigation will be reported.

McDonald *et al.* (1976) reported an experiment carried out in Saskatoon, Canada, where picloram was applied at 1.12 kg $ha^{-1}$ and 2.24 kg $ha^{-1}$ to a bromegrass (*Bromus inermis*)–alfalfa (*Medicago sativa*) field (coarse-textured, low organic-matter soil) prior to sprinkler irrigation. Two water regimes were imposed, one being the natural rainfall and the other the natural rainfall augmented by sprinkler irrigation. The rainfall received on the field plots was 14.9 cm while the irrigated plots received a total of 27.4 cm of water. Figure 9 shows the comparative distribution of picloram and water in soil in the two treatments. The wetting and drying cycles induced a reconcentration of picloram in the upper layer of the soil after its initial displacement with depth. The redistribution of picloram with depth was affected by the initial amount of herbicides applied to the soil surface before being irrigated. We should emphasize that under sprinkler irrigation we are dealing with one-dimensional transport so that only the vertical distribution of picloram is of importance. The behavior of picloram in field experiments was the same as observed in column experiments of Davidson and Chang (1972) and Davidson and McDougal (1973) on the transport of the chemical related to the properties of the soil and pore water velocity.

The fluctuation in soil water content in an irrigated field also affects the pathway of picloram degradation in soil. Guenzi and Beard (1976) in a

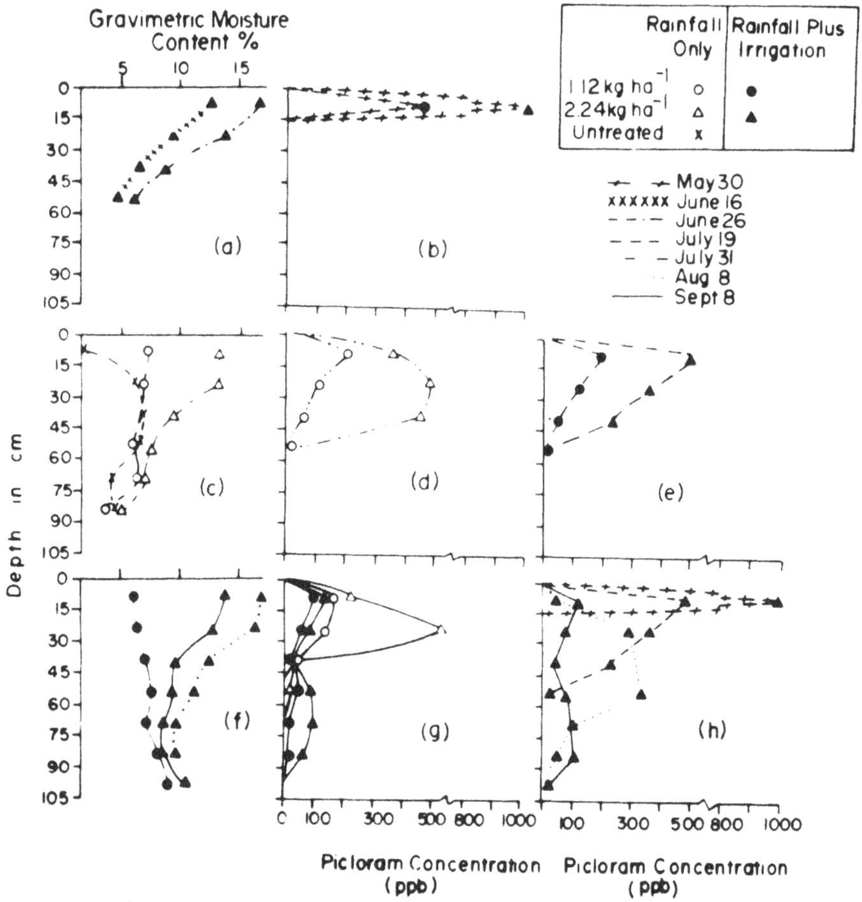


**Figure 9.** Soil moisture contents (a, c, f) and picloram distribution in the soil (b, d, e, g, h) at various times through the growing season. Figure a shows soil moisture content and b the herbicide concentration shortly after herbicide application. Figures d and e give the distribution patterns of the herbicide midway through the growing season. Figures c and f show the moisture contents towards the end of the growing season and gives the final herbicide distributions observed, while h shows the changes in herbicide distribution throughout the season. From McDonald *et al.*, *Soil Sci.* 121: 94–102. Copyright 1976, The Williams & Wilkins Co., Baltimore.

laboratory experiment showed that picloram degradation rates during 20-day incubation periods at field capacity, interrupted by 16-day drying cycles, varied among soils and decreased in general after each successive drying cycle at 30 and 50°C. Figure 10 shows the picloram degradation rates as a function of soil water content and temperature from five soils, when soils were dried at a constant rate from field capacity to 15 bars tension. The degradation rate decreased gradually as water content decreased from field capacity (0.33 bar) to 15 bars tension and ceased after

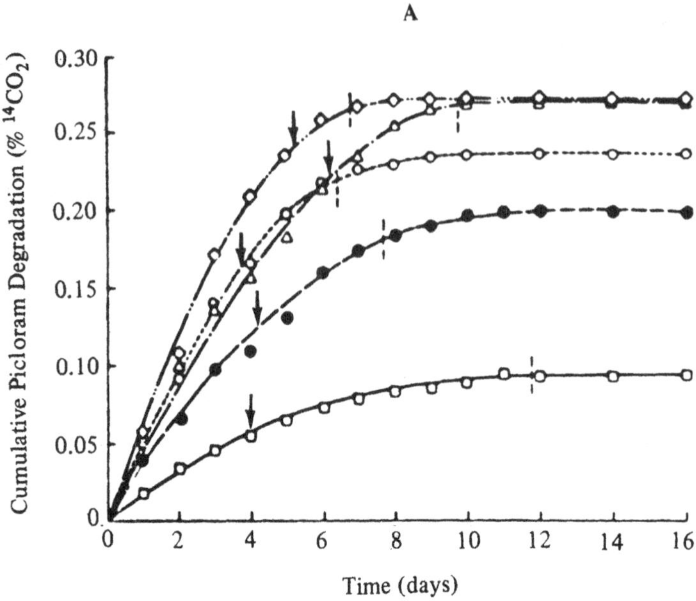


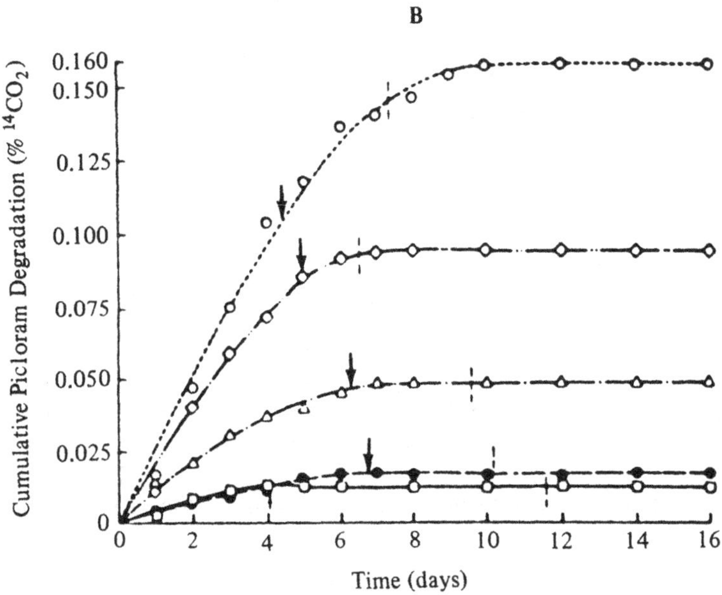


**Figure 10.** Effect of temperature (A = 50°C, B = 30°C) on picloram degradation in soils. Arrows indicate water content at 15 bars tension; dashed vertical lines indicate approximately one monolayer of water on soil surfaces. Reproduced from Guenzi and Beard, *J. Environ. Qual.* 5: 189–192 (1976), by permission of the Am. Soc. of Agronomy, Crop Sci. Soc. of America, Soil Sci. Soc. of America.

the soils were air dried. From this information, it may be concluded that not only the dispersion of a herbicide in soil is affected by irrigation, but also its degradation rate.

The effect of trickle irrigation on the distribution of herbicides in soil—applied before the onset of irrigation—has been reported by Koren and Shlevin (1977). They used simazine and GS-14259 in a simulated drip irrigation experiment carried out on a red Mediterranean soil (96% sand; 0.5% OM) and a drip irrigation experiment in the field on a loessial soil (17% clay; 0.4% OM). The herbicides were applied to dry soil 24 hrs before water was applied using tricklers. Redistribution of herbicides in the soil was determined by bioassays on samples taken at various times after irrigation. Figure 11 shows the two-dimensional distribution of simazine and GS-14259 in a loessial soil 15 and 30 hrs after start of drip irrigation.

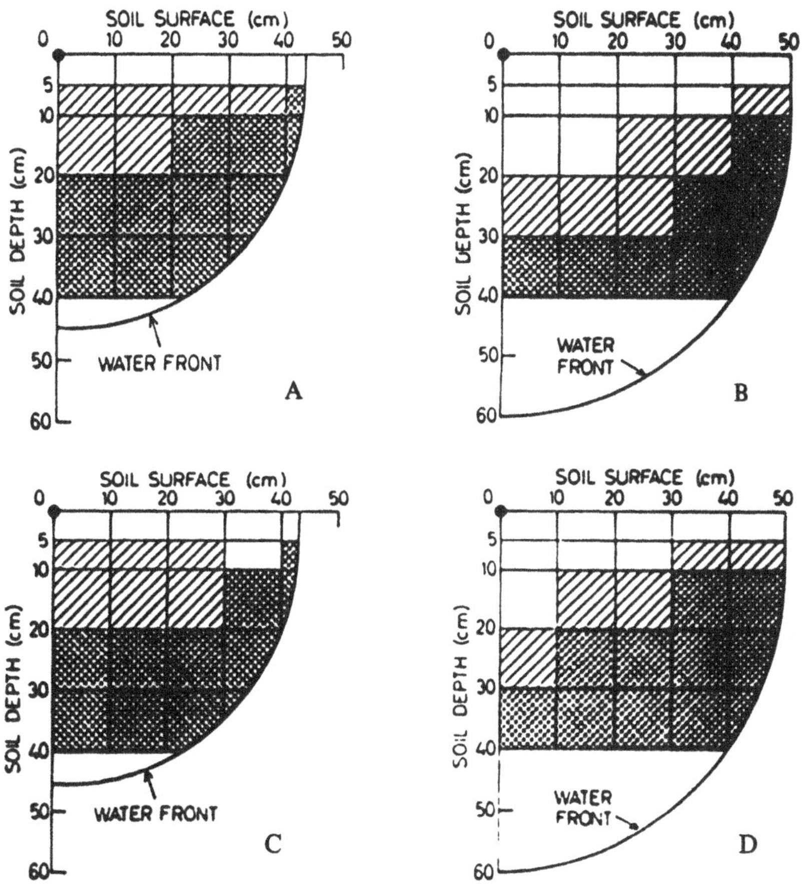


**Figure 11.** Distribution of simazine (A and B) and GS-14259 (C and D) in a loessial soil after 15 (A and C) and 30 (B and D) hrs of drip irrigation. From Koren and Shlevin (1977).

The data of Koren and Shlevin (1977) show that the herbicides are leached around the trickle nozzle, but the leaching pattern is defined by the properties of the herbicide and the amount of water applied. In the experiment on loessial soil the herbicide movement lags behind the water front displacement. The retardation of simazine is less than that of GS-14259. In the red Mediterranean sandy soil, simazine was completely leached from around the trickler and concentrated at the bottom of the wetted zone.

The effect of irrigation on the residual activity of herbicides was the object of a series of experiments. Lange *et al.* (1976), for example, evaluated effectiveness of several herbicides (napropamide, pebulate, chloramben) for annual weed control in processing tomatoes grown under drip and furrow irrigation on a Panoche clay loam, a fine loamy, mixed (calcareous), Thermic Typic Torriorthent. A continuous wetting along the drip line appeared to enhance degradation or leaching of the herbicide resulting in lower degree of control of pigweed and barnyard grass, compared to furrow irrigation. Jennings and Drennan (1979) proved the potential effectiveness of trifluralin and oxadiazon applied as preplant herbicides in an irrigated cotton field in Sudan. Incorporation of the herbicides 6 months before planting was as effective as a cleanweeding by hand.

One practice of applying herbicide prior to irrigation includes spraying the banks of dry canals to control canal weeds. A problem that arises by this practice is that the herbicide subsequently may be found in the irrigation water. Reed canary grass (*Phalaris arundinacea*) is one of the main weeds infesting irrigation canals. In two separate experiments, the residues of TCA and glyphosate were determined in the irrigation water following treatments of the banks of dry canals (Comes *et al.*, 1975, 1976). TCA was detected in the irrigation water entering the canal 5 months after application. The concentration was highest during the first hour of water flow and disappeared after 4 hrs. No detectable residues of glyphosate or its soil metabolite, aminomethylphosphonic acid, were detected in the water entering the channels about 5 months after application. The soil adsorptive capacity for the two herbicides and their persistence in soil were responsible for the low residual concentration found in the irrigation water.

A herbicide applied prior to irrigation may be dissipated by runoff from the treated area. Despite the importance of irrigation-induced runoff, little attention was given to this problem until recent years. Mielke and Leavitt (1981) studied the amount of atrazine and alachlor lost in runoff water and sediment under center pivot irrigation. They found that the water solubility of a herbicide affected the runoff pattern. Herbicides that are water soluble were more susceptible to initial runoff events, and a greater portion of the total amount lost was dissolved in water. Yet the amount of herbicide carried out on the sediment was minimal even with a relatively insoluble herbicide such as atrazine. The accumulative herbicide loss owing to runoff after three consecutive irrigations did not exceed 0.1% of the amount applied even for the most favorable conditions for runoff. The method of irrigation had a marked effect on the amount of herbicide lost. Low pressure spray induced an average runoff of 0.052% for atrazine and 0.074% for alachlor; high pressure system did not induce runoff of alachlor and induced only 0.001% runoff of atrazine.

Herbicide losses in runoff water associated with furrow irrigation was measured recently by Spencer *et al.* (in press). Seasonal losses of the soil-applied herbicides (cycloate, DCPA, dinitramine, prometryn, and trifluralin) were usually 1 to 2% of the amounts applied. Concentration was highest in the first irrigation following herbicide application. Time elapsed between pesticide application and the irrigation event was inversely related to log concentration indicating an approximately first-order rate of decrease in runoff concentration with time. The concentration of prometryn and DCPA in runoff water vs. time after application is presented in Figure 12. Spencer *et al.* (in press) found that the concentrations and the amounts of herbicides in the 0 to 1-cm depth of soil proved to be good indicators of the concentrations and amounts of pesticides to be expected in runoff water under furrow irrigation systems.

### C. Herbigation

The use of irrigation systems as herbicide carriers increased in the last decade, and a new term, ***herbigation***, was introduced to define the practice of applying chemicals to soil using irrigation water. There is little scientific information in the literature about herbigation, and the existing papers are mainly devoted to technological aspects of usage, e.g., advantages and disadvantages, uniformity of distribution, development of antipollution devices, etc. However, some research on herbigation has been published by Crutchfield and Wiese (1977), Crutchfield and Wiese (1978), Fischbach and Martin (1975, 1976), Ogg (1980), Smith *et al.* (1973), and Wiese and Turner (1978).

Application of herbicides via irrigation requires an advanced irrigation technology able to control the amount of water applied per unit of land. Pressure irrigation systems (sprinkler or trickle) are the most convenient ones for herbicide incorporation into soil via irrigation water. Water has been used as a herbicide carrier under furrow or flood irrigation also, but to a lesser extent. One risk of applying herbicides through irrigation water is water pollution from back flow into the irrigation system, a risk that may be prevented by proper irrigation management and by the installation of proper safety devices.

Distribution of herbicides in soil resulting from their incorporation via sprinkler and drip irrigation will be illustrated by presenting selected examples from the literature.

The effect of sprinkler irrigation systems—specifically center pivot—on

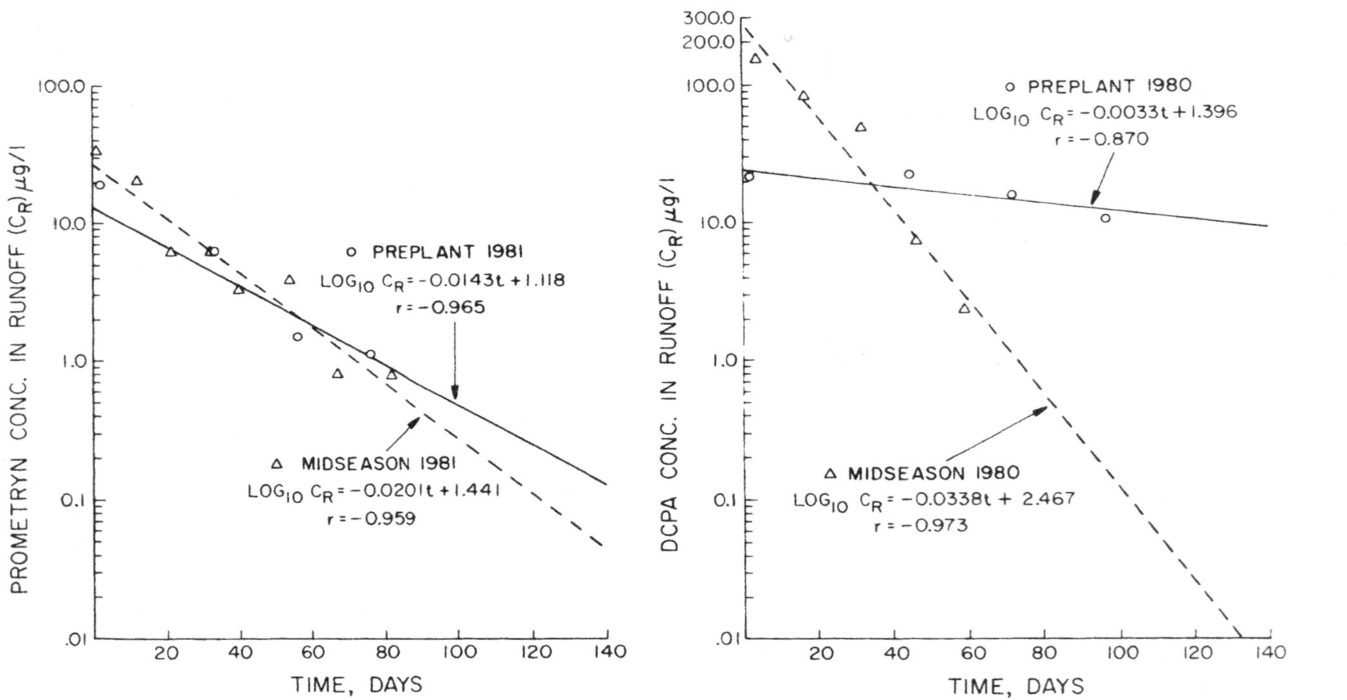


**Figure 12.** Relationship between prometryn and DCPA concentration in runoff water and time elapsed since their preplant or midseason application to field I. After Spencer (in press).

herbicide distribution and soil penetration was determined in the U. S. coastal plains by Dowler *et al.* (1982). Butylate, metholachlor, alachlor, pendimethalin, oryzalin, metribuzin alone or in combination with atrazine, and/or simazine were applied to corn, peanuts, and soybeans. The experiment was carried out on nine soils characterized by a clay content ranging between 3.6 and 4.2% and organic matter from 0.6 to 2.18%. The authors showed that the penetration and distribution in soil of the herbicides were affected by the properties of the compounds and of the soil, seed bed preparation, and water application rate. Contrary to expectations based on water solubility, metholachlor moved very little in the soil profile while butylate was found in substantial concentrations in the soil profile between 10 and 30-cm depth. Results also indicated that water application rate and seedbed preparation technique affected herbicide distribution when applied with irrigation water. Figure 13 shows the effect of seedbed preparation on herbicide distribution in soil when applied by sprinkler irrigation. The tandem disc-harrow favored, in general, a retention of herbicides in the second upper layer—compared to subsoil plant seedbed preparation.

Application of herbicides via sprinkler irrigation may cause harmful biocidal effects on the irrigated crop. Bruns *et al.* (1976) reported on an experiment where fenac was applied by sprinkler irrigation to sugarbeets, potatoes (*Solanum tuberosum*), dwarf maize, grain sorghum (*Sorghum vulgare*), soybeans, and field beans (*Phaseolus vulgaris*) growing on Warden very fine sandy loam, a coarse silty, mixed, Mesic Xerollic Cambortheds (1.5% OM). Injury to foliage of sprinkler-irrigated potatoes, soybeans, and field beans increased as the rate of fenac increased. However, the yield was reduced only on plots treated with a total of 0.24 kg/ha.

Another aspect to be considered in application of herbicides via sprinkler irrigation is the potential losses of the chemical before reaching the land surface. An experiment carried out in Texas (Crutchfield *et al.*, 1977) showed that when atrazine was applied through sprinklers 10% was lost, but 85% of the trifluralin disappeared between the time it left sprinkler nozzles and the time it landed on the soil surface. Photochemical decomposition and volatilization probably explains the herbicide losses. Consequently, the application of herbicides via sprinkler irrigation is a questionable practice when the chemical is characterized by a high vapor pressure or is photochemically unstable.

Improved drip irrigation technology offers a new approach for a safer and more efficient method of introducing herbicides into the soil environment. A study of the movement of several herbicides applied via drip systems was reported by Nygen and Lange (1978), who found that oryzalin, EPTC, and trifluralin moved farther from the emitter than did

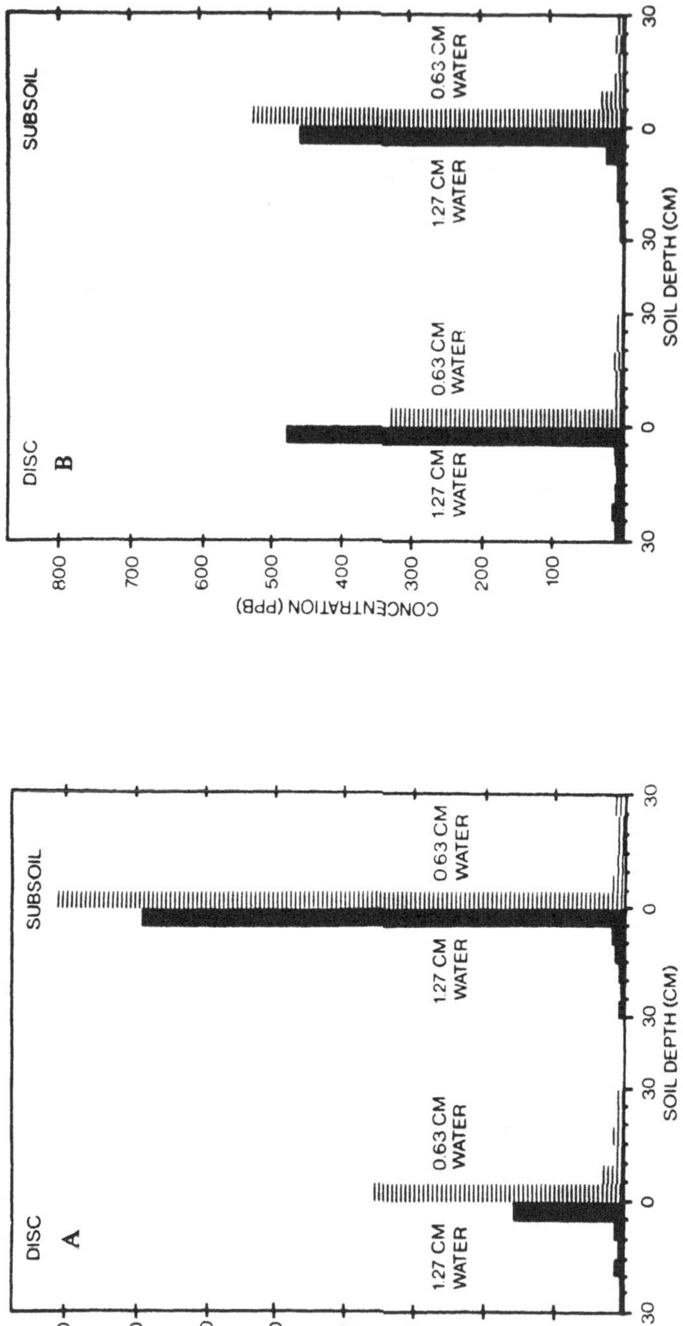


**Figure 13.** Effect of irrigation water rates and seedbed tillage on metachlor (A) and oryzalin (B) in sandy soil. After Dowler *et al.* (1982).

napropamide, pebulate, or chloramben. Lange and Schlesselman (1977) reported relative movement of drip-applied herbicides as napropamide > oryzalin > trifluralin > EPTC, but in all cases herbicide movement was considerably less than water movement.

Recently Gerstl *et al.* (1981), Gerstl and Yaron (1983b), and Gerstl and Albasel (1984) measured distribution of bromacil and napropamide in soils when applied by drip irrigation. The experiment was carried out on four soils with various physico-chemical and mineralogical properties. The parameters studied were water application rate (4.0 and 1.5 L/hr) and the effect of wetting and drying cycles. In general, napropamide, which was moderately adsorbed by the soils, was concentrated around the emitter; at the high water application rate the compound moved laterally more than vertically and it penetrated to a greater depth at the lower application rate. Bromacil, which was weakly adsorbed by the soils, was more uniformly distributed in the soil but did exhibit slight retardation in the heavier soils. Figure 14 shows the effect of soil type on two-dimensional distribution of the herbicides and water as affected by properties of the soil. Soil structure affects the distribution of applied chemicals. Figure 15 shows the distribution of water, napropamide, and bromacil applied with drip irrigation in an aggregated soil. The soil structure greatly influenced the distribution of the applied chemical resulting in deeper penetration than might otherwise be expected. This phenomenon may be explained only by a bypass flow of the solute.

The redistribution of bromacil and napropamide in an irrigated tomato field following their application through a drip-irrigation system was studied by Gerstl and Albasel (1984). Effect of rate (2 L/hr and 4 L/hr) and frequency of irrigation (daily and weekly) was tested. Irrigation rate and frequency had pronounced effects on the distribution of these herbicides in soil. The concentration of bromacil and napropamide at 20-cm distance from the dripper as affected by rate and frequency of irrigation are presented in Tables 9 and 10. Loss of bromacil 6 weeks after application by herbigation may be explained by leaching; however, the disappearance of napropamide was attributed to volatilization and photochemical decomposition as well as leaching. Gerstl and Albasel (1984) also demonstrated the effect of soil spatial variability and crop effect on napropamide distribution in the soil. Table 11 shows the standard deviation of soil moisture content (PW%) and napropamide (N, mg $kg^{-1}$) from three preplanted cropped plots irrigated daily at 4 L $hr^{-1}$ and sampled 8 days after herbicide application. It may be observed that the standard deviation of the napropamide content is much larger than the SD of the soil water content. The fact may be explained by the heterogeneity of soil properties that affect the behavior of interacting chemicals such as napropamide.

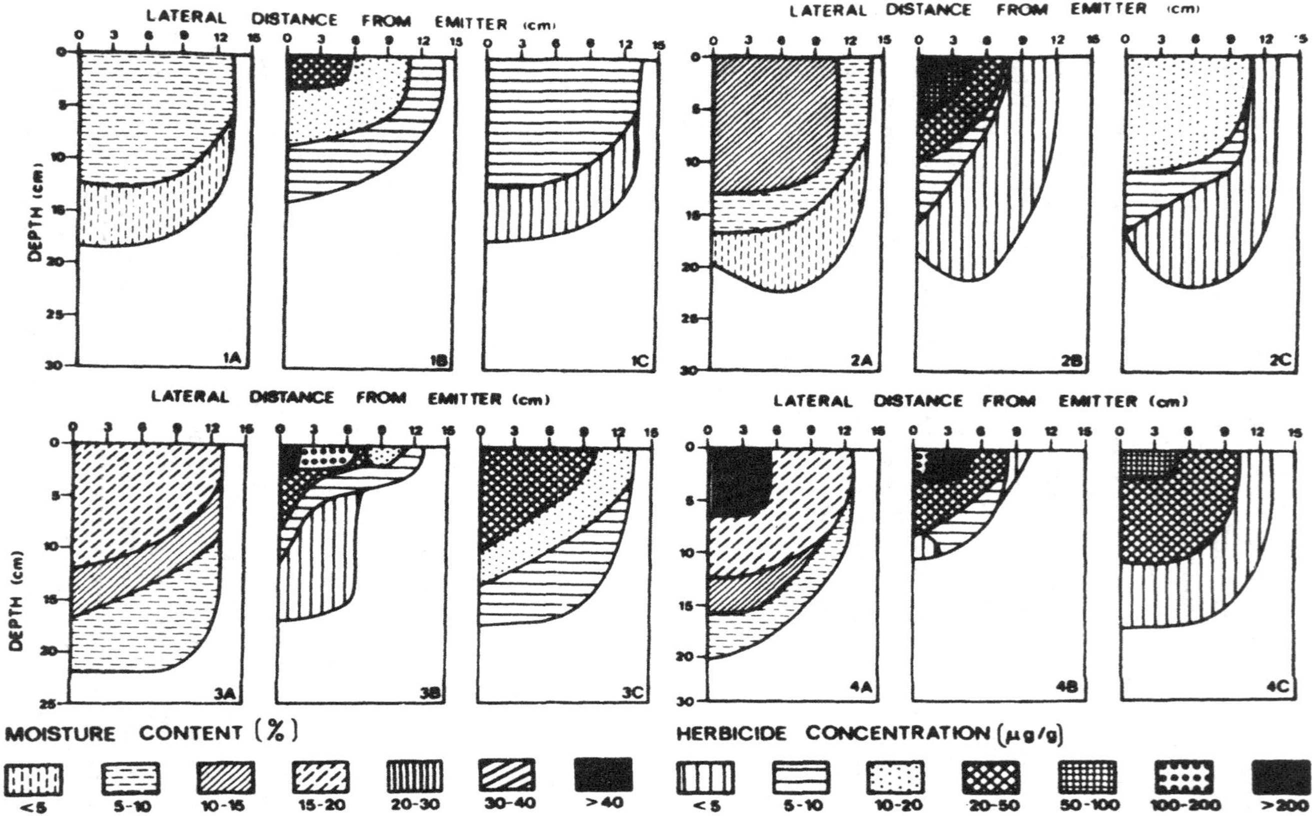


**Figure 14.** Effect of soil type on the distribution of (A) water, (B) napropamide, and (C) bromacil 24 hrs after application of the herbicide solution at 4 L/hr to the (1) Mivtachim, (2) Bet-Dagan 1, (3) Gilat, and (4) Bet-Dagan II soils. Reproduced from Gerstl and Yaron *Soil Sci. Soc. Am. J.* 47:474–478 (1983b), by permission of the Soil Sci. Soc. of America.

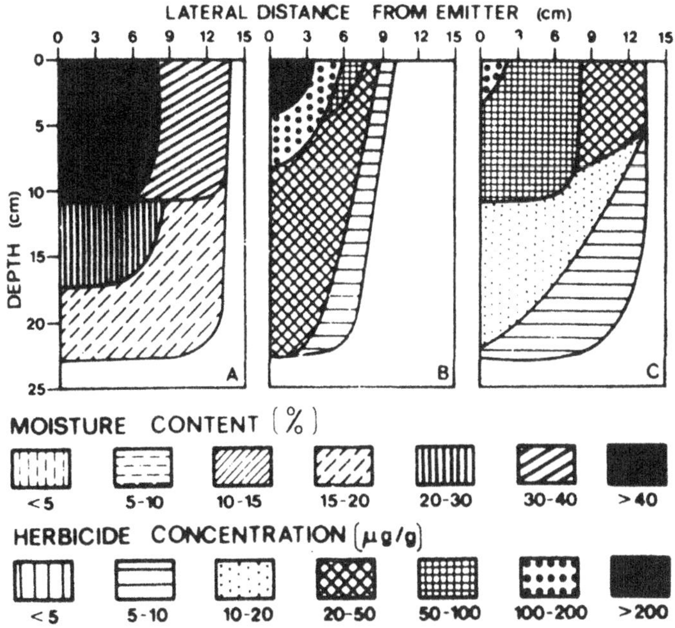


**Figure 15.** Distribution of (A) water, (B) napropamide, and (C) bromacil 24 hrs after application of a herbicide solution to an aggregated clay soil at 4 L/hr. Reproduced from Gerstl and Yaron *Soil Sci. Soc. Am. J.* 47:474–478 (1983b), by permission of the Soil Sci. Soc. of America.

**Table 9.** Effect of Irrigation Rate on the Distribution of Bromacil (B) and Napropamide (N) in a Red Mediterranean Soil[a]

| | Flow rate 2 L/hr | | | | Flow rate 4 L/hr | | | |
|---|---|---|---|---|---|---|---|---|
| Time (days) | 7 | | 42 | | 7 | | 42 | |
| Depth (cm) | N | B | N | B | N | B | N | B |
| | mg/kg | | | | | | | |
| 0–5 | 15.0 | 17.3 | 1.0 | 1.0 | 55.0 | 15.0 | 1.2 | 2.3 |
| 5–10 | 30.0 | 5.2 | 0.2 | 0.3 | 11.0 | 9.5 | 0.4 | 0.3 |
| 10–20 | 30.0 | 1.6 | 0.1 | 0 | 4.7 | 12.5 | 0.9 | 1.2 |
| 20–40 | 5.7 | 5.8 | 0.2 | 0.1 | 3.7 | 8.6 | <0.1 | 0 |
| 40–60 | 1.2 | 10.3 | 0.4 | 0.1 | 3.2 | 7.0 | <0.1 | 0 |
| 60–80 | 0.7 | 1.8 | 0.2 | <0.1 | 1.4 | 1.2 | <0.1 | <0.1 |
| 80–1000 | 2.7 | 2.5 | 0.2 | <0.1 | 0.3 | 0.6 | 0 | 0 |

[a]After Gerstl and Albasel (1984); sampled at 20 cm distance from the emitter.

**Table 10.** Effect of Irrigation Frequency on the Distribution of Bromacil (B) and Napropamide (N) in a Red Mediterranean Soil[a]

| Time (days) | Daily | | | | | | Weekly | | | | | |
|---|---|---|---|---|---|---|---|---|---|---|---|---|
| | 7 | | 28 | | 42 | | 7 | | 21 | | 36 | |
| Depth (cm) | N | B | N | B | N | B | N | B | N | B | N | B |
| | mg/kg | | | | | | | | | | | |
| *0–5* | *55.0* | *15.0* | *5.4* | *0.8* | *1.2* | *2.3* | *47.0* | *3.6* | *21.7* | *1.9* | *1.7* | *0.2* |
| 5–10 | 11.0 | 9.5 | 4.1 | 0.3 | 0.4 | 0.3 | 32.0 | 2.8 | 17.3 | 0.6 | 1.5 | 0.2 |
| 10–20 | 4.7 | 12.5 | 2.4 | 0.1 | 0.9 | 1.2 | 16.0 | 1.4 | 11.7 | 1.3 | 0.9 | 0.1 |
| 20–40 | 3.7 | 8.6 | 1.7 | 0.5 | 0.1 | 0 | 8.5 | 3.3 | 6.2 | 0.5 | 1.0 | 0.1 |
| 40–60 | 3.2 | 7.0 | 1.4 | 0.9 | 0.1 | 0 | 1.3 | 2.1 | 3.0 | 1.0 | 0.4 | 0.2 |
| 60–80 | 1.4 | 1.2 | 0.9 | 1.5 | 0.1 | 0.1 | 1.8 | 3.0 | 6.2 | 1.5 | 0.5 | 0.5 |
| 80–100 | 0.3 | 0.6 | 0.6 | 0.9 | 0 | 0.3 | 1.1 | 0.6 | 3.1 | 0.5 | 0.5 | 0.8 |

[a] After Gerstl and Albasel (1984); sampled at 20 cm distance from the emitter; flow rate 4L/hr.

**Table 11.** Average ± Standard Deviation of Soil Moisture Content (PW, %) and Napropamide (N, mg $kg^{-1}$) from Three Replicated Cropped Plots Irrigated Daily at 4 L $hr^{-}$ and Sampled 8 Days After Herbicide Application[a]

| Depth (cm) | Radial distance from dripper (cm) 0 | | 10 | | 20 | | 30 | |
|---|---|---|---|---|---|---|---|---|
| | PW | N | PW | N | PW | N | PW | N |
| 0–5 | 20.1 ± 0.6 | 7.0 ± 1.1 | 19.0 ± 0.9 | 14.1 ± 11.5 | 16.4 ± 1.1 | 22.7 ± 32.4 | 12.8 ± 1.4 | 0.3 ± 0.3 |
| 5–10 | 19.4 ± 1.2 | 10.4 ± 8.5 | 19.3 ± 1.1 | 8.8 ± 1.6 | 18.2 ± 2.0 | 10.7 ± 14.3 | | |
| 10–20 | 18.9 ± 2.3 | 9.4 ± 3.7 | 19.7 ± 1.1 | 5.2 ± 2.6 | 19.4 ± 2.2 | 3.3 ± 3.5 | 14.9 ± 3.3 | 0.0 ± 0.0 |
| 20–40 | 18.9 ± 2.3 | 5.2 ± 0.8 | 18.2 ± 2.2 | 3.3 ± 2.2 | 17.6 ± 1.5 | 0.9 ± 0.9 | 16.1 ± 1.9 | 0.4 ± 0.3 |
| 40–60 | 20.4 ± 2.1 | 1.8 ± 1.0 | 18.4 ± 2.0 | 2.5 ± 2.8 | 19.9 ± 2.1 | 0.5 ± 0.6 | | |

[a] After Gerstl and Albasel (1984).

# VI. Conclusions

Irrigation can assist in optimizing herbicide behavior in the soil medium and in the unsaturated zone or zones between the land surface and ground water. As previously discussed, processes governing the fate of herbicides in soil, such as adsorption and volatilization, are affected by soil moisture content from the very moment that the chemical reaches the soil. Competition between water and chemicals, specifically non-polar herbicides, for soil adsorption sites may lead to an increase in herbicide availability when applied to a prewetted soil. Likewise, in the case of herbicides with high vapor pressures, their application on wet soil may unduly enhance their volatility. By irrigation, we are able to control the soil moisture content to optimize the efficiency of soil-applied herbicides by reducing their adsorption on soil surfaces or by decreasing their volatilization.

Fluctuation of soil moisture content during irrigation changes the soil–water–air ratio, which controls the ratio between aerobic and anaerobic conditions and consequently the soil biological activity. This involves changes in the anaerobic-aerobic status of the whole soil profile or microsite differences. An example of microsite difference is aerobic conditions on the external face of an aggregate and anaerobic conditions in the interior of the same aggregate. Since the degradation of herbicides is mainly biologically induced, fluctuations in aerobic and anaerobic conditions affect their rates of degradation. By changing frequency or timing or irrigations with respect to herbicide applications, we can control somewhat the aerobic-anaerobic conditions and as a result, the kinetics of biological degradation of a herbicide incorporated into the soil.

Undoubtedly, flow rate and amount of water applied will affect the leaching of herbicides and their three-dimensional distribution in soil or other porous media. It is within our capability in irrigated agriculture to control those factors by various irrigation techniques and thereby to prevent excessive leaching of applied herbicides from the soil surface into deeper horizons and their loss from the soil surface by runoff. Prevention of ground-water pollution through leaching of herbicides into soil and surface water pollution by preventing irrigation-induced runoff may be achieved by controlling flow rate and amount of water applied on irrigated land.

Even though the properties of a herbicide are the main factor in deciding its fate in the soil environment, it is possible by proper management of soil water to increase the herbicide's efficiency and thereby decrease the amount applied per unit of surface, and to minimize its transport below the rootzone, thereby preventing leaching of the chemical into ground water. Irrigation therefore becomes a tool in the control of herbicide behavior in the soil environment.

In assessing the behavior of herbicides in soil in general, and under irrigation in particular, we are still using parameters obtained at equilibrium, under saturated conditions, and in uniform homogeneous soils. Natural fields display a wide variation of their chemical physico-chemical, and hydraulic properties, making highly questionable the frequent practice of treating such a field as homogeneous soil. Irrigation practices add to the spatial heterogeneity of the soil properties by creating saturated and non-saturated conditions in the soil.

Changes in soil moisture status and their effects on herbicides in irrigated soils lead to non-equilibrium situations in the field. It is natural that in such a situation discrepancy often is observed between the behavior of herbicides in the field and their predicted behavior based on laboratory obtained parameters. There is a need for new experimental procedure by which the adsorption and behavior of herbicides in unsaturated conditions can be measured experimentally.

Soil transport properties measured on a field scale are spatially variable with orders of magnitude difference in properties occurring in relatively small fields. The ability to formulate and numerically solve complex situation models for the behavior of herbicides on a field scale using the data base presently available is not yet possible. Consequently, new research should be devoted to determining parameters able to take into account field-scale heterogeneity to reduce uncertainty of the model predictions at the field scale level.

Reliable predictions to be obtained by the above recommended research will result in increased efficiency of herbicide application and decreased ground-water pollution under irrigation.

## Acknowledgment

The support of United States–Israel Binational Agricultural Research and Development Fund (BARD) is acknowledged.

**Appendix.** Common Name, Chemical Name, Solubility (ppm), and Vapor Pressure (mm Hg) of Herbicides Mentioned in the Text.

| Common name | Chemical name | Solubility (ppm)[a] | Vapor pressure (mm hg)[a,b] |
|---|---|---|---|
| 1. *Aliphatics* | | | |
| Acrolein | 2-propenal | – | 210 (20) |
| Cacodylic acid | dimethylarsinic acid | 668,000 | – |
| Dalapon | 2,2-dichloropropionic acid | 900,000 | – |
| DSMA | disodium methanearsonic acid | 279,000 | – |
| Glyphosate | *N*-(phosphonomethyl) glycide | 12,000 | – |
| MAA | methanearsonic acid | – | – |
| MAMA | monoammonium methanearsonic acid | – | – |
| MSMA | monosodium methanearsonic acid | 1,400,000 | – |
| Methyl bromide | methyl bromide | 13,400 | – gas – |
| TCA | trichloroacetic acid | 10,000,000 | del. |
| 2. *Amides* | | | |
| Alachlor | 2-chloro-*N*-(2,6-diethylphenyl)-*N*-methoxymethyl acetamide | 242 | $2.2 \times 10^{-5}$ (25) |
| Benzamide | | | |
| Butachlor | *N*-(butoxymethyl)-2-chloro-*N*-(2,6-diethylphenyl) acetamide | 23 | $4.5 \times 10^{-6}$ (25) |
| Diphenamid | *N,N*-dimethyl-2,2-diphenylacetamide | 260 | – |
| Flamprop | *N*-benzoyl-*N*-(3-chloro-4 fluorophenyl)-D-alaninate | 10 | $2.37 \times 10^{-5}$ (20) |
| Metolachlor | 2-chloro-*NO*(2-ethyl-6 methylphenyl)-*N*-(2-methoxy-l-methylethyl) acetamide | 530 | $1.3 \times 10^{-5}$ (20) |
| Napropamide | 2-(α-naphthoxy)-*N,N*-diethylpropionamide | 73 | $4 \times 10^{-5}$ (20) |
| Pronamide | *N*-(1,1-dimethylpropynl)-3,5-dichlorobenzamide | 15 | $8.5 \times 10^{-5}$ |
| Propanil | *N*-(3,4-dichlorophenyl) propionamide | 225 | $9 \times 10^{-5}$ (60) |
| R-25788 | *N,N*-diallyl-2,2-dichloroacetamide | 5000 | $6 \times 10^{-3}$ (25) |

| | | | |
|---|---|---|---|
| 3. *Benzoics* | | | |
| Amiben (Chloramben) | 3-amino-2,5-dichlorobenzoic acid | 700 | $7.0 \times 10^{-3}$ (100) |
| Dicamba | 3,6-dichloro-O-anisic acid | 4,500 | $3.41 \times 10^{-5}$ (25) |
| 2,3,6-TBA | 2,3,6-trichlorobenzoic acid | 7,700 | $2.4 \times 10^{-2}$ (100) |
| 4. *Benzonitrites* | | | |
| Bromoxynil | 3,5-dibromo-4-hydroxybenzonitrile | 130 | – |
| Dichlorobenil | 2,6-dichlorobenzonitrile | 21 | $5.5 \times 10^{-4}$ (20) |
| 5. *Bipyridyliams* | | | |
| Diquat | 6,7-dihydrodipyrido [1,2-α2′, 1′-c] pyrazinediium ion | 700,000 | $9.8 \times 10^{-8}$ |
| Paraquat | 1,1′-dimethyl-4,4′-bipyridinium ion | vs | – |
| 6. *Dinitroanilines* | | | |
| Benefin | *N*-butyl-*N*-ethyl-α,α,α-trifluoro-2,6-dinitro-*p*-toluidine | 0.1 | $7.8 \times 10^{-5}$ (25) |
| Butralin | 4-(1,1-dimethylethyl)-*N*-(1-methylpropyl)-2,6-dinitro benzeneamine | 1 | $1.3 \times 10^{-5}$ (25) |
| Dicryl (chloroanocryl) | *N*-(3,4-dichlorophenyl)-2-methyl-2-propenamide | | |
| Dinitramine | *N*,*N*³-diethyl 2,4-dinitro-6-trifluoromethyl-1,3-benzeneamine | 1 | $3.6 \times 10^{-6}$ (25) |
| Fuchloralin | *N*-(2-chloroethyl)-α,α,α-trifluoro-2,6-dinitro-*N*-propyl-*p*-tuluidine | <1 | $2.8 \times 10^{-5}$ (20) |
| Isopropalin | 2,6-dinitro-*N*, *N*-dipropylcumidine | 0.08 | $3 \times 10^{-5}$ (26) |
| Oryzaline | 3,5-dinitro-*N*⁴,*N*⁴-dipropylsulfanilamide | 2.5 | $<1 \times 10^{-7}$ (30) |
| Pendimethalin | *N*-(1-ethylpropyl )-3,4 dimethyl-2,6 dinitrobenzeneamine | 0.5 | $3 \times 10^{-5}$ (25) |
| Profluralin | *N*-(cyclopropylmethyl)-α,α,α-trifluoro-2,6-*N*-propyl-*p*-toluidine | 0.1 | $6.9 \times 10^{-5}$ (20) |
| Trifluralin | α,α,α-trifluoro-2,6-dinitro-*N*-*N*-dipropyl-*p*-toluidine | 0.3 | $1.1 \times 10^{-4}$ (25) |

**Appendix** *(Continued)*

| Common name | Chemical name | Solubility (ppm)[a] | Vapor pressure (mm hg)[a,b] |
|---|---|---|---|
| 7. *Phenoxys* | | | |
| 2,4-D | (2,4-dichlorophenoxy)acetic acid | 900 | 0.4 (160) |
| 2,4-DB | 4-(2,4-dichlorophenoxy)butyric acid | 46 | – |
| Dichlorprop (2,4-DP) | 2-(2,4-dichlorophenoxy)propionic acid | 710 | – |
| MCPA (phenothiol) | 2-methyl-4-chlorophenoxyacetic acid | 825 | $1.5 \times 10^{-6}$ (21) |
| 2,4,5-T | (2,4,5-trichlorophenoxy) acetic acid | 156 | $5.3 \times 10^{-9}$ (25) |
| 8. *Carbamates* | | | |
| Asulam | methyl sulfanilylcarbamate | 5,000 | – |
| Barban | 4-chloro-2-butynyl *N*-(3-chlorophenyl) carbamate | 11 | $3.8 \times 10^{-7}$ (25) |
| Benthiocarb | (*S*)-(4-chlorobenzyl)*N*,*N*-diethylthiolcarbamate | 30 | $1.3 \times 10^{-6}$ (20) |
| Butylate | (*S*)-ethyl dinobutylthiocarbamate | 45 | $1.3 \times 10^{-4}$ (25) |
| Chlorpropham | isopropyl-*m*-chlorocarbanitate | 88 | $\sim 10^{-5}$ (25) |
| Cycloate | (*S*)-ethyl *N*-ethylthiocyclohexanecarbamate | 85 | $6.2 \times 10^{-3}$ (25) |
| EPTC | (*S*)-ethyl dipropylthiocarbamate | 370 | $3.4 \times 10^{-4}$ (25) |
| Methan (Vapam) | methyldithiocarbamic acid | 722,000 | – |
| Molinate | (*S*)-ethyl hexhydro-1 H-azepine-1-carbothioate | 800 | $5.6 \times 10^{-3}$ (25) |
| Pebulate | (*S*)-propyl butylethylthiocarbamate | 60 | $3.5 \times 10^{-4}$ (25) |
| Phenmedipham | methylam-hydroxycarbamate *m*-methylcarbanilate | 47 | $9.8 \times 10^{-9}$ |
| Pirimicarb | 2-(dimethylamino)-t,6-dimethyl-4-pyrimilinyl dimethyl carbonate | 2,700 | $3 \times 10^{-5}$ (30) |
| Propham (IPC) | isopropyl carbanilate | 250 | - sub - |
| Swep | methyl 3,4-dichlorophenylcarbamate | – | – |

| | | | |
|---|---|---|---|
| 9. *Diphenylesters* | | | |
| Acifluorfen | 5-[2-chloro-4-(trifluoromethyl)phenoxy]-2-nitrobenzoic acid | 120 | $10^{-6}$ (20) |
| Bifenox | methyl 5-(2,4-dichlorophenoxy)-2-nitrobenzoate | 0.35 | $2.4 \times 10^{-4}$ (30) |
| Chlomethoxyfen | 4-(2,4-dichlorophenoxy)-2-methoxy-1-nitrobenzene | 0.3 | |
| Nitrofen | 2,4-dichlorophenyl *p*-nitrophenyl ether | 1 | $8 \times 10^{-6}$ (40) |
| 10. *Triazines* | | | |
| Atrazine | 2-chloro-4-(ethylamino)-6-(isopropylamino)-*s*-triazine | 33 | $3 \times 10^{-7}$ (20) |
| Credazine | 3-(2-methylphenoxy)pyridazine | 2,000 | – |
| Cyanazine | 2-chloro-4-(1-cyano-1-methylethylamino)-6-ethylamino-*s*-triazine | 165 | $1.6 \times 10^{-9}$ (20) |
| Metribuzin | 4-amino-6-tert-butyl-3-(methylthio)-*s*-triazin-5(4H)-one | 1,220 | $1 \times 10^{-5}$ (20) |
| Prometon | 2,4-bis(isopropylamino)-6-methoxy-*s*-triazine | 750 | $2.3 \times 10^{-6}$ (20) |
| Prometryn | 2,4-bis (isopropylamino)-6-(methylthio)-*s*-triazine | 48 | $1 \times 10^{-6}$ (20) |
| Propazine | 2-chloro-4,6-bis(isopropylamino)-*s*-triazine | 8.6 | $2.9 \times 10^{-8}$ (20) |
| Simazine | 2-chloro-4,6-bis (ethylamino)-*s*-triazine | 3.5 | $6.1 \times 10^{-9}$ (20) |
| Simetryne | 2,4-bis(ethylamino)-6-(methylthio)-*s*-triazine | 450 | – |
| Terbumeton (GS14259) | 2-(tert-butylamino)-4-(ethylamino)-6-methoxy-*s*-triazine | 130 | $2 \times 10^{-6}$ (20) |
| Terbuthylazine | 2-(tert-butylamino)-4-chloro-6-(ethylamino)-*s*-triazine | 8.5 | $1.1 \times 10^{-6}$ (20) |
| Terbutryne | 2-(tert-butylamino)-4-ethylamino)-6-(methylthio)-*s*-triazine | 25 | $9.6 \times 10^{-6}$ (20) |
| 11. *Uracils* | | | |
| Bromacil | 5-bromo-3-sec-butyl-6-methyluracil | 815 | $2.4 \times 10^{-7}$ |
| Lenacil | 3-cyclohexyl-5,6-trimethyleneuracil | 6 | – |
| Terbacil | 3-tert-butyl-5-chloro-6-methyluracil | 710 | $4.8 \times 10^{-7}$ (30) |

**Appendix** *(Continued)*

| Common name | Chemical name | Solubility (ppm)[a] | Vapor pressure (mm hg)[a,b] |
|---|---|---|---|
| 12. *Substituted Ureas* | | | |
| Chlorsulfuron | 2-chloro-*N*-[(4-methoxy-6-methyl-1,3,5-triazin-2-yl) aminocarbonyl]-benzenesulfonamide | 0.03–2.8 | $4.6 \times 10^{-6}$ (25) |
| Chlortoluron | *N'*-(3-chloro-4-methylphenyl)-*N,N*-dimethylurea | 70 | $1.3 \times 10^{-7}$ (20) |
| Diuron | 3-(3,4-dichlorophenyl)-1,1-dimethylurea | 42 | $3.1 \times 10^{-6}$ (50) |
| Fenuron | *N,N*-dimethyl-*N'*-phenylurea | 3,850 | $1.6 \times 10^{-4}$ (60) |
| Fluometuron | 1,1-dimethyl-3-(α,α,α-trifluoro-*m*-toly)urea | 90 | $5 \times 10^{-7}$ (20) |
| Isoproturon | *N,N*-dimethyl-*N'*-[4-(1-methylethyl)phenyl]urea | 55 | $2.5 \times 10^{-8}$ (20) |
| Linuron | *N'*-(3,4-dichlorophenyl)-*N*-methoxy-*N*-methylurea | 75 | $1.5 \times 10^{-5}$ (24) |
| Methabenzthiazuron | *N*-2-benzothiazoly-*N,N'*-dimethylurea | 59 | $10^{-6}$ (20) |
| Monuron | *N'*-(4-chlorophenyl)-*N,N*-dimethylurea | 230 | $5 \times 10^{-6}$ (25) |
| Tebuthiuron | *N*-[5-(1,1-dimethylethyl)-1,3,4-thiadiazol-2-yl]-*N,N'*-dimethylurea | 2,300 | $2 \times 10^{-6}$ (25) |
| 13. *Others* | | | |
| Amitrole | 3-amino-*s*-triazole | 280,000 | |
| Bentazon | 3-isopropyl-1H-2,1,3-benzothiadiazin-4(3H)-one 2,2-dioxide | 500 | $<7.5 \times 10^{-11}$ |
| Buthidazole | 1-(5-t-butyl-1,3,-thiadiazol-2-yl)-3-methyl-5-hydroxy-1,3-imidazolidin-2-one | 4,900 | $4 \times 10^{-8}$ |
| Chloridazon | 5-amino-4-chloro-2-phenyl-3(2H)-pyridazinone | 400 | $<7.5 \times 10^{-8}$ |
| Dacthal (DCPA) | dimethyl tetrachloroterephthalate | 0.5 | >0.01 (40) |
| Dinoseb (DNBP) | 2-(1-methylpropyl)-4,6-dinitrophenol | 52 | 1(151) |
| DNOC | 2-methyl-4,6-dinitrophenol | 130 | $1 \times 10^{-4}$ (25) |
| Endothall | 3,6-endoxohexahydrophthalic acid | 100,000 | – |

| | | | |
|---|---|---|---|
| Fenac | (2,3,6-trichlorophenyl)acetic acid | 200 | $8.3 \times 10^{-3}$ (100) |
| Oxadiazon | 2-tert-butyl-4-(2,4-dichloro-5-isopropoxyphenyl)-$\Delta^2$-1,3,4-oxadiazolin-5-one | 0.7 | $10^{-6}$ (20) |
| Penthachlophenol | Pentachlorophenol | 20 | 0.12 (100 |
| Picloram | 4-amino-3,5,6-trichloropiclinic acid | 430 | $6.2 \times 10^{-7}$ (35) |

[a]Solubility and vapor pressure data taken from Pesticide Manual (1983) and Herbicide Handbook of the Weed Science Society of America (1983).
[b]Numbers in parentheses are temperature at which vapor pressures were determined.

## References

Abernathy, J.R., and J.M. Davidson. 1971. Effect of calcium chloride on prometryne and fluometuron adsorption in soil. *Weed Sci.* 19:517–521.

Adams, R.S., Jr. 1972. Effect of soil organic matter on the movement and activity of pesticides in the environment. In: D.D. Hemphill (ed.), *Trace substances in environmental health.* V. Proc. Conf. Univ. Missouri. University of Missouri.

Adams, R.S., Jr., Jr. 1973. Factors influencing soil adsorption and bioactivity of pesticides. *Res. Rev.* 47:1–54.

Armstrong, D.E., and G. Chesters. 1968. Adsorption catalyzed chemical hydrolysis of atrazine. *Environ. Sci. Tech.* 2:683–689.

Armstrong, D.E., and J.G. Konrad. 1974. Nonbiological degradation of pesticides. In: W.G. Guenzi (ed.), *Pesticides in soil and water.* Soil Science Society of America, Madison. pp. 123–131.

Arnold, J.S., and W.J. Farmer. 1979. Exchangeable cations and picloram sorption by soil and model adsorbents. *Weed Sci.* 27:257–262.

Ashton, F.M., and A.S. Crafts. 1973. *Mode of action of herbicides.* John Wiley and Sons, New York.

Bailey, G.W., and J.L. White. 1970. Factors influencing the adsorption, desorption, and movement of pesticides in soil. *Res. Rev.* 32:29–92.

Bailey, G.W. and J.L. White (1965). Herbicides: A compilation of their physical, chemical and biological properties. Residue Reviews 10, 97–112.

Bailey, G.W., J.L. White, and T. Rothberg. 1968. Adsorption of organic herbicides by montmorillonite: role of pH and chemical character of adsorbate. *Soil Sci. Soc. Am. Proc.* 32:222–234.

Bartha, R. 1971. Fate of herbicide-derived chloroanilines in soil. *J. Agr. Food Chem.* 19:385–387.

Baumaister, P., and K. Hurle. 1975. Abbav von Atrazine und Simazine unter Freilandbedingungen. *Z. Plf. Krankh. Pflschulz* 7:154–167.

Bear, J. 1972. *Dynamics of fluids in porous media.* American Elsevier, New York.

Beckett, P.H.T., and R. Webster. 1971. Soil variability: a review. *Soils Fert.* 34:1–15.

Best, J.A., and J.B. Weber. 1979. Disappearance of *s*-triazines as affected by soil pH using a balance sheet approach. *Weed Sci.* 22:364–373.

Biggar, J.W., and D.R. Nielsen . 1976. Spatial variability of leaching characteristics of a field soil. *Water Resour. Res.* 12:78–84.

Bouma, J. 1981. Soil morphology and preferential flow along macropores. *Agr. Water Manage.* 3:235–250.

Bovey, R.W., E. Burnett, C. Richardson, J.R. Baur, M.G. Merkle, and D.E. Kissel. 1975. Occurrence of 2,4,5-T and picloram in subsurface water in the blacklands of Texas. *J. Environ. Qual.* 4:103–106.

Bowman, B.T. 1981. Anomalies in the log Freundlich equation resulting in deviations in adsorption K values of pesticides and other organic compounds when the system of units is changed. *J. Environ. Sci. Health* B16:113–123.

Bowman, B.T., and W.W. Sans. 1979. The aqueous solubility of twenty-seven insecticides and related compounds. *J. Environ. Sci. Health* B14:625–534.

Bresler, E. 1971. The effect of irrigation method on soil water regime. In: *Bull. Rech. Agr. de Gembloux Hors. Serie*. 233–254.

Bresler, E., G. Dagan, R.J. Wagenet, and A. Laufer. 1984. Statistical analysis of salinity and texture: effects on spatial variability of soil hydraulic conductivity. *Soil Sci. Soc. Am. J.* 48:16–25.

Briggs, G.G. 1981. Theoretical and experimental relationships between soil adsorption, octanol-water partition coefficients, water solubilities, bioconcentration factors, and the parachlor. *J. Agr. Food Chem.* 29:1050–1059.

Brock, T.D. 1970. *Biology of microorganisms*. Prentice-Hall, Englewood Cliffs, NJ.

Brown, C.B., and J.L. White. 1969. Reaction of 12 *s*-triazines with soil clays. *Soil Sci. Soc. Am. Proc.* 33:863–867.

Bruns, V.F., J.M. Hodgson, A.D. Kelley, and R.D. Comes. 1976. Responses and residues in certain field crops irrigated with water containing fenac herbicide residues. *Bull. Wash. State Univ. Coll. Agr. Res. Cent.* July 1976, 828. 8 p.

Burgess, T.M., and R. Webster. 1980. Optimal interpolation isarithmic mapping of soil properties. I. The semivariogram and punctual kriging. *J. Soil Sci.* 13:315–331.

Burns, R.G. 1978. Enzyme activity in soil: some theoretical and practical considerations. In: R.G. Burns (ed.), *Soil enzymes*. Academic Press, New York. pp. 295–340.

Burns, R.G., M.H.B. Hayes, and M. Stacey. 1973a. Some physico-chemical interactions of paraquat with soil organic materials and model compounds. I. Effects of temperature, time, and adsorbate degradation on paraquat adsorption. *Weed Res.* 13:67–78.

Burns, R.G., M.H.B. Hayes, and M. Stacey, M. 1973b. Some physico-chemical interactions of paraquat with soil organic materials and model compounds. II. Adsorption and desorption equilibria in aqueous suspensions. *Weed Res.* 13:79–90.

Calvet, R. 1980. Adsorption-desorption phenomena. In: R.J. Hance (ed.), *Interactions between herbicides and the soil*. Academic Press, London. pp. 1–30.

Calvet, R. 1983. Behavior of pesticides in unsaturated zone. Adsorption and transport phenomenon. In: B. Yaron, G. Dagan, and J. Goldschmid (eds.), *Pollutants in porous media*. Springer-Verlag, Heidelberg. pp. 143–151.

Caverley, D.J. 1978. Establishment of crop damage levels in soils for the herbicides lenacil, linuron, and trifluralin. *Proc. 1978 Brit. Prot. Conf. Weeds*, 549–555.

Chen, Y.L. 1983. Behavior and fate of pesticides in paddy ecosystems. In: J. Miyamoto and P.C. Kearney (eds.), *Pesticide chemistry: human welfare and the environment. Vol. 2*. Pergamon Press, Oxford. pp. 339–344.

Chiska, H., and P.C. Kearney. 1970. Metabolism of propanil in soils. *J. Agr. Food Chem.* 18:854–858.

Clay, D.V., and K.G. Scott. 1973. The persistence and penetration of large doses of simazine in uncropped soil. *Weed Res.* 13:42–50.

Cliath, M.M., W.F. Spencer, W.J. Farmer, T.D. Shoup, and R. Grover. 1980. Volatilization of *S*-ethyl *N*, *N*-dipropylthiocarbamate from water and wet soil during and after flood irrigation of an alfalfa field. *J. Agr. Food Chem.* 28:610–613.

Combellack, J.H. 1984. Herbicide application: a review of ground application techniques. *Crop Prot.* 3:9–35.

Comes, R.D., V.F. Bruns, and A.D. Kelley. 1976. Residues and persistence of glyphosphate in irrigation water. *Weed Sci.* 24:47–50.

Comes, R.D., P.A. Frank, and R.J. Demint. 1975. TCA in irrigation water after bank treatments for weed control. *Weed Sci.* 23:207–210.

Crank, J. 1975. *Mathematics of diffusion*. 2nd ed. Clarendon Press, Oxford.

Crosby, D.G. 1983. The fate of herbicides in California rice culture. In: T. Miyamoto and P.C. Kearney (eds.), *Pesticide chemistry: human welfare and environment*. Vol. 2. Pergamon Press, Oxford. pp. 339–346.

Crosby, D.G., and M.Y. Li. 1969. Herbicide photodecomposition. In: P.C. Kearney and D.D. Kaufman (eds.), *Degradation of herbicides*. Marcel Dekker, New York. pp. 321–363.

Crosby, D.G., K.W. Moilonen, and J.N. Seiber. 1978. Pesticide degradation by atmospheric oxidants. *Abstr. Fourth Intl. Congr. Pest. Chem.* Zurich.

Crutchfield, D.A., and A.F. Wiese. 1978. Herbicide deposition and loss with sprinkler irrigation. *Weed Sci. Soc. Am. Abstr.* No. 5.

Crutchfield, D.A., R.N. Clark, and A.P. Wiese. 1977. Deposition of atrazine and trifluralin with sprinkler irrigation. *Proc. 30th Ann. Mtng. South. Weed Sci. Soc.*. p. 428.

Cussans, G.W., S.R. Moss, R.J. Hance, S.J. Embling, D. Caverly, T.G. Marks, and J.J. Palmer. 1982. The effect of tillage method and soil factors on the performance of chlortoluron and isoproturon. *Proc. 1982 Brit. Crop Prot. Conf.* 1:153–160.

Dagan, G., and E. Bresler. 1979. Solute dispersion in unsaturated heterogeneous soil at field scale. II. Applications. *Soil Sci. Soc. Am. J.* 43:467–472.

Dagan, G., and E. Bresler. 1984. Solute transport in soil at field scale. In: B. Yaron,

G. Dagan, and J. Goldschmid (eds.), *Pollutants in porous media.* Springer-Verlag, Heidelberg. pp. 17–26.

Dao, T.H., T.L. Lavy, and R.C. Sorensen. 1979. Atrazine degradation and residue distribution in soil. *Soil Sci. Soc. Am. J.* 43:1129–1134.

Davidson, J.M., and R.K. Chang. 1972. Transport of picloram in relation to soil physical conditions and pore-water velocity. *Soil Sci. Soc. Am. Proc.* 36:257–261.

Davidson, J.M., and J.R. McDougal. 1973. Experimental and predicted movement of three herbicides in a water-saturated soil. *J. Environ. Qual.* 2:428–433.

Davidson, J.M., G.H. Brusewitz, D.R. Baker, and A.L. Wood. 1975. *Use of soil parameters for describing pesticide movement through soils.* EPA-660/2-75-009.

Davison, J.G., and D.V. Clay. 1972. The persistence of some horticultural herbicides in soil. *Rep. Weed Res. Org.* 1970–71, pp. 58–65.

Demint, R., J. Frank, A. Peter, and R.D. Comes. 1970. Amitrole residues and rate of dissipation in irrigation water. *Weed Sci.* 18:439–442.

Deuel, L.E., F.T. Turner, K.W. Brown, and J.D. Price. 1977. Persistence and factors affecting dissipation of molinate under flooded rice culture. *J. Environ. Qual.* 7:373–377.

Dowler, C., W.A. Rhode, L.E. Fetzer, D.E. Scott, Sr., T.E. Sklany, and C.W. Swann. 1982. The effect of sprinkler irrigation on herbicide efficacy, distribution, and penetration in some coastal plain soils. *Univ. Georgia Agr. Exp. Sta. Res. Bull.* No. 281. 28 pp.

Doyle, R.C., D.D. Kaufman, and G.W. Burt. 1978. Effect of dairy manure and sewage sludge on $^{14}$-C-pesticide degradation in soil. *J. Agr. Food Chem.* 18:854–858.

Edwards, C.A. 1972. Insecticides. In: C.A.I. Goring and J.W. Hamaker (eds.), *Organic chemicals in the soil environment. Vol. 1.* Marcel Dekker, New York. pp. 513–568.

Elesparu, R., W. Lijinsky, and J.K. Setlow. 1974. Nitrosocarbaryl as a potent mutagan of environmental significance. *Nature* 247:386–387.

Farmer, W.J., and Y. Aochi. 1974. Picloram sorption by soil. *Soil Sci. Soc. Am. Proc.* 38:418–423.

Faust, S.D., and I.H. Suffet. 1966. Recovery, separation and identification of organic pesticides from natural and potable waters. *Res. Rev.* 15:44–116.

Felsot, A., and P.A. Dahm. 1979. Sorption of organophosphorous and carbamate insecticides by soil. *J. Agr. Food Chem.* 27:557–559.

Finkel, H.J. 1982. *Handbook of irrigation technology* . CRC Press, Boca Raton, FL.

Fischbach, P.E., and A. Martin. 1975. *Herbigation—applying herbicide through sprinkler systems.* Am. Soc. Agr. Eng. Paper No. 75-1527.

Fischbach, P.E., and A. Martin. 1976. Application of herbicides through a center pivot sprinkler. *Weed Sci. Soc. Am. Abst.* No. 247.

Fuhr, G., and W. Mittelstaedt. 1980. Plant experiments on the bioavailability of unextracted (carboxyl $^{14}C$) methabenzthiazuron residues from soil. *J. Agr. Food Chem.* 26:605–607.

Gerstl, Z., and N. Albasel. 1984. Field distribution of pesticides applied via a drip irrigation system. *Irrig. Sci.* 181–193.

Gerstl, Z., and U. Mingelgrin. 1979. A note on the adsorption of organic molecules on clays. *Clays Clay Miner.* 27:285–290.

Gerstl, Z., and U. Mingelgrin. 1984. Sorption of organic substances by soils and sediments. *J. Environ. Sci. Health* B19:297–312.

Gerstl, Z., and B. Yaron. 1977. Adsorption and desorption of parathion by attapulgite as afftected by the mineral structure. *J. Agr. Food Chem.* 26:569–573.

Gerstl, Z., and B. Yaron. 1983a. Behavior of bromacil and napropamide in soils: I. Adsorption and degradation. *Soil Sci. Soc. Am. J.* 47:474–478.

Gerstl, Z., and B. Yaron. 1983b. Behavior of bromacil and napropamide in soils. II. Distribution after application from a point source. *Soil Sci. Soc. Am. J.* 47:478–483.

Gerstl, Z., U. Mingelgrin, and B. Yaron. 1977. Behavior of vapam and methylisothiocyanate in soils. *Soil Sci. Soc. Am. J.* 41:545–548.

Gerstl, Z., P.H. Nye, and B. Yaron. 1979a. Diffusion of a biodegradable pesticide. II. As affected by microbial decomposition. *Soil Sci. Soc. Am. J.* 43:843–848.

Gerstl, Z., B. Yaron, and P.H. Nye. 1979b. Diffusion of a biodegradable pesticide. I. In a biologically inactive soil. *Soil Sci. Soc. Am. J.* 43:839–842.

Gerstl, Z., S. Saltzman, L. Kliger, and B. Yaron. 1981. Distribution of herbicides in soil in a simulated drip irrigation system. *Irrig. Sci.* 2:155–166.

Getzin, L.V., and R.K. Chapman. 1960. The fate of phoratein soils. *J. Econ. Entomol.* 53:47–51.

Glotfelty, D.E., A.W. Taylor, B.C. Turner, and W.H. Zoller. 1984. Volatilization of surface-applied pesticides from fallow soil. *J. Agr. Food Chem.* 32:638–643.

Goring, C.A.I., and J.W. Hamaker (edes.). 1972. *Organic chemicals in the soil environment. Vol. I.* Marcel Dekker, New York.

Goring, C.A.I., D.A. Laskowski, J.W. Hamaker, and R.W. Meikle. 1975. Principles of pesticide degradation in soils. In: R. Haque and V.H. Freed (eds.), *Environmental dynamics of pesticides.* Plenum Press, New York. pp. 135–172.

Gray, T.R.G. 1970. Microbial growth in soils. In: G.F. Guyer (ed.), *Pesticides in the soil: ecology degradation, and movement.* Michigan State University, East Lansing, MI. pp. 36–41.

Grim, R.E. 1968. *Clay mineralogy.* 2nd Ed. McGraw-Hill, New York.

Grover, R., S.R. Shewchuk, A.J. Cessna, A.E. Smith, and J.H. Hunter. In press. Fate of 2,4-D iso-octyl ester after application to a wheat field, a mass balance approach. *J. Environ. Qual.*

Guenzi, W.D. (ed.). 1974. *Pesticides in soil and water.* Soil Science Society of America, Madison.

Guenzi, W.D., and W.E. Beard. 1974. Volatilization of pesticides. In: W.E. Guenzi (ed.), *Pesticides in soil and water.* Soil Science Society of America. Madison. pp. 108–122.

Guenzi, W.D., and W.E. Beard. 1976. Picloram degradation in soils as influenced by soil water content and temperature. *J. Environ. Qual.* 5:189–192.

Guth, J.A. 1980. The study of transformations. In: R.J. Hance (ed.), *Interaction between herbicides and the soil.* Academic Press, New York. pp. 123–157.

Haider, K. 1983. Anaerobic microsites in soils and their possible effect on pesticide degradation. In: J. Miyamoto and P.C. Kearny (eds.), *Pesticide chemistry: human welfare and the environment. Vol. III.* Pergamon Press, Oxford. pp. 351–356.

Hall, J.K., M. Pawlus, and E.R. Higgins. 1972. Losses of atrazine in runoff water and soil sediments. *J. Environ. Qual.* 1:172–176.

Hamaker, J.W. 1972a. Decomposition: quantitative aspects. In: C.A.I. Goring and J.W. Hamaker (eds.), *Organic chemicals in the soil environment. Vol. I.* Marcel Dekker, New York. pp. 253–340.

Hamaker, J.W. 1972b. Diffusion and volatilization. In: C.A.I. Goring and J.W. Hamaker (eds.), *Organic chemicals in the soil environment. Vol. I.* Marcel Dekker, New York. pp. 341–397.

Hamaker, J.W., and J.M. Thompson. 1972. Adsorption. In: C.A.I. Goring and J.W. Hamaker (eds.), *Organic chemicals in the soil environment. Vol. I.* Marcel Dekker, New York. pp. 49–144.

Hamaker, J.W., C.A.I. Goring, and C.R. Youngson. 1966. Sorption and leaching of 4 amino -3,5,6-trichloropicolinic acid in soil. *Adv. Chem. Ser.* 60:23–37.

Hamaker, J.W., C.R. Youngson, and C.A.I. Goring. 1967. Prediction of the persistence and activity of Tordon herbicide in soils under field conditions. *Down to Earth* 23:30–36.

Hance, R.J. 1965. The adsorption of urea and some of its derivatives by a variety of soils. *Weed Res.* 5:98–107.

Hance, R.J. 1969. An empirical relationship between chemical structure and the sorption of some herbicides by soils. *J. Agr. Food Chem.* 17:667–668.

Hance, R.J. 1970. Influence of sorption on the decomposition of pesticides. *Soc. Chem. Ind. Monogr.* 37:32–104.

Hance, R.J. 1973. The effect of nutrients on the decomposition of the herbicides atrazine and linuron incubated with soil. *Pest. Sci.* 4:817–822.

Hance, R.J. 1979. Effect of pH on the degradation of atrazine, dichlorprop, linuron, and propyzamide in soil. *Pest. Sci.* 10:83–86.

Hance, R.J. (ed.). 1980a. *Interactions between herbicides and the soil.* Academic Press, London.

Hance, R.J. 1980b. Transport in the vapour phase. In: R.J. Hance (ed.), *Interactions of herbicides with soil.* Academic Press, London. pp. 59–81.

Hance, R.J., and C.E. McKone. 1971. Effect of concentration on the decomposition rates in soil of atrazine, linuron and picloram. *Pest. Sci.* 3:31–34.

Hanks, R.J., and G.L. Ashcroft. 1980. *Applied soil physics.* Springer-Verlag, Heidelberg.

Hanks, R.J., H.R. Gardner, and M.L. Fairbourn. 1967. Evaporation of water from soils as influenced by drying with wind or radiation. *Soil Sci. Soc. Am. Proc.* 31:593–598.

Haque, R., and V.H. Freed (eds.). 1975. *Environmental dynamics of pesticides.* Plenum Press, New York.

Harper, L.A., A.W. White, Jr., R.R. Bruce, A.W. Thomas, and R.A. Leonard. 1976. Soil and microclimate effects on trifluralin volatilization. *J. Environ. Qual.* 5:236–242.

Harris, C.I. 1966. Adsorption, movement and phytotoxicity of monuron and *s*-triazine herbicides. *Weeds* 14:6–10.

Harris, C.I., and C.F. Warren. 1964. Adsorption and desorption of herbicides by soil. *Weeds* 12:120–126.

Harris, C.I., E.A. Woolson, and B.E. Hummer. 1969. Dissipation of herbicides at three soil depths. *Weed Sci.* 17:27–31.

Harris, C.R., R.A. Chapman, C. Harris, and C.M. Tu. 1984. Biodegradation of pesticides in soil: rapid induction of carbamate degrading factors after carbofuran treatment. *J. Environ. Sci. Health* B19:1–11.

Harris, J.C. 1982. Rate of hydrolysis. In: W.J. Lyman, W.F. Reehl, and D.H. Rosenblatt (eds.), *Handbook of chemical property estimation methods.* McGraw-Hill Book Co., New York. pp. 7-1–7-48.

Hartley, G.S., and I.J. Graham-Bryce. 1980. *Physical principles of pesticide behaviour.* Academic Press, London.

Hassan, A. 1982. Report on first FAO/IAEA research coordination meeting on isotopictracer-aided studies of bound pesticide residues in soil, plants, and food. *J. Environ. Sci. Health* B17:291–295.

Hasset, J.J., J.C. Means, W.L. Wanwart, and S.G. Wood. 1980. *Sorption properties of sediments and energy-related pollutants.* EPA-600/3-80-041.

Hayes, M.H.B. 1970. Adsorbtion of triazine herbicides on soil organic matter, including a short review on soil organic chemistry. *Res. Rev.* 57:1–25.

Helling, C.S. 1971. Pesticide mobility in soils. II. Application of soil thin-layer chromatography. *Soil Sci. Soc. Am. Proc.* 35:737–743.

Helling, C.S., and A.E. Krivonak. 1978. Physicochemical characteristics of bound dinitroaniline herbicides in soils. *J. Agr. Food Chem.* 26:1156–1164.

Herbicide Handbook, 5th Edition, 1983. *Weed Science Society of America.* p. 515.

Hill, I.R. 1978. Microbial transformations of pesticide. In: I.R. Hill and S.J.L. Wright (eds.), *Pesticide microbiology.* Academic Press, New York. pp. 137–202.

Hiltibran, R.C., and A.J. Turgeon. 1977. Creeping bentgrass response to aquatic herbicides in irrigation water. *J. Environ. Qual.* 6:263–267.

Hilton, H.W., and Q.H. Yuen. 1963. Adsorption of several preemergence herbicides by Hawaiian sugarcane soils. *J. Agr. Food Chem.* 11:230–243.

Hornsby, A.G., and J.M. Davidson. 1973. Solution and adsorbed fluometuron concentration distribution in a water saturated soil. Experimental and predicted evaluation. *Soil Sci. Soc. Am. Proc.* 37:823–828.

Hubbert, M.K. 1956. Darcy's law and the field equations of the flow of underground fluids. *Trans. Am. Inst. Min. Met. Eng.* 207:222–239.

Hurle, K. 1982. Unterschungen zum abbau von herbiziden in boden. *Acta Phytomedica* 8:1–120.

Hurle, K., and A. Walker. 1980. Persistence and its prediction. In: R.J. Hance (ed.), *Interactions between herbicides and the soil.* Academic Press, London. pp. 83–122.

Ishikawa, K., U. Asano, and Y. Nakamura. 1976. Behavior and disappearance of benthiocarb herbicide in water, soil and rice plants in paddy fieldes treated with its granular formulations. *Weed Res., Japan* 21:16–21.

Ishikawa, K., R. Shinohara, A. Yari, S. Shigematsu, and I. Kimura. 1980. Identification of *S*-benzyl *N,N*-diethyl-thiocarbamate in paddy field soils applied with benthiocarb herbicide. *J. Pest. Sci.* 5:107–109.

Jennings, E.A., and D.S.H. Drennan. 1979. Weed control with pre-sowing herbicides in irrigated cotton (*Gossypium harbadense* L.) and groundnut (*Arachis hypogaea* L.) in the Sudan. II. Economic, labour requirement, crop safety and operational aspects of trifluralin and benfluralin. *1979 Eur. Weed Res. Soc. Weed Res.* 19:301–306.

Jensen, M.E. (ed.). 1980. *Design and operation of farm irrigation systems.* American Society of Agricultural Engineers, St. Joseph, MI.

Jury, W.A. 1983. Chemical transport modeling: current approaches and unresolved problems. In: *Chemical mobility ad reactivity in soil systems.* Soil Sci. Soc. Am. Spec. Publ. No. 11. Soil Science Society America, Madison. pp. 49–64.

Jury, W.A., W.J. Farmer, and W.F. Spencer . 1984a. Behavior assessment model for trace organics in soil. II. Chemical classification and parameter sensitivity. *J. Environ. Qual.* 13:567–572.

Jury, W.A., W.F. Spencer, and W.J. Farmer. 1983a. Behavior assessment model for trace organics in soil. I. Model description. *J. Environ. Qual.* 12:558–564.

Jury, W.A., W.F. Spencer, and W.J. Farmer. 1983b. Use of models for assessing relative volatility, mobility, and persistence of pesticides and other trade organics in soil systems. In: J. Saxena (ed.), *Hazard assessment of chemicals current developments. Vol. 2.* Academic Press, New York. pp. 2–43.

Jury, W.A., W.F. Spencer, and W.J. Farmer. 1984b. Behavior assessment model for trace organics in soil. III. Application of screening model. *J. Environ. Qual.* 13:573–579.

Jury, W.A., R. Grover, W.F. Spencer, and W.J. Farmer. 1980. Modeling vapor losses of soil-incorporated triallate. *Soil Sci. Soc. Am. J.* 44:445–450.

Kaufman, D.D., Y. Katan, D.A. Edwards, and E.G. Jordan. In press. Microbial adaptation and metabolism of pesticides. In: J.L. Hilton (ed.), *Agricultural chemicals of the future. Vol. 8.* Beltsville Symp. in Agr. Res.

Kearny, P.C., and D.D. Kaufman. 1969. *Degradation of herbicides.* Marcel Dekker, New York.

Khan, S.U. 1979. Adsorption of bipyridylium herbicides by humic acid. *J. Environ. Qual.* 3:202–206.

Khan, S.U. 1980. *Pesticides in the soil environment.* Elsevier Scientific Publishing Co., Amsterdam.

Khan, S.U., and H.A. Hamilton. 1980. Extractable and bound (nonextractable) residues of prometryn and its metabolites in an organic soil. *J. Agr. Food Chem.* 28:126–132.

Khan, S.U., and K.C. Ivanson. 1981. Microbiological released unextracted bound residues from an organic soil treated with prometryn. *J. Agr. Food Chem.* 29:1301–1303.

Klingman, G.C., and F.M. Ashton. 1982. *Weed science.* 2nd ed. John Wiley and Sons, New York.

Knisel, W.G. (ed.). 1980. *CREAMS: a field-scale model for chemicals, runoff, and erosion, from agricultural management systems.* USDA Cons. Res. Rep. 26. USDA, Washington, D.C.

Koren, E., and E. Shlevin. 1977. Effect of trickle irrigation on soil-applied herbicides. *Weed Res.* 17:133–138.

Kuwatsuka, S. 1983. Fate of herbicides in flooded paddy soils. In: T. Miyamoto and P.C. Kearney (eds.), *Pesticide chemistry: human welfare and environment. Vol. 2.* Pergamon Press, Oxford. pp. 347–354.

Lange, A., and S. Schlesselman. 1977. The effect of injecting four herbicides through plastic irrigation emitters on the control of annual grasses. *West. Soc. Weed Sci. Res. Prog. Rep.* p. 45.

Lange, A., H. Kempe, H. Fischer, and J. Schlesselman. 1976. Chemical weed control with drip irrigation. *Proc. West. Soc. Weed Sci.* 29:53–58.

Lavy, T.L. 1970. Diffusion of three chloro *s*-triazines in soil. *Weed Sci.* 18:53–56.

Leistra, M. 1973. The use of computation models in studies on herbicide behavior in soil. *Proc. Env. Weed Res. Coun. Herb.-Soil.* pp. 216–226.

Leistra, M. 1977. A model for the transport of pesticides in soil with diffusion control rates of adsorption and desorption. *Agr. Environ.* 3:325–335.

Leistra, M. 1978. Computed restribution of pesticides in the root zone on an arable crop. *Plant and Soil* 49:569–580.

Leistra, M. 1980. Transport in solution. In: R.J. Hance (ed.), *Interactions between herbicides and the soil.* Academic Press, New York, pp. 31–58.

Leistra, M., J.H. Smelt, and R. Zandvoort. 1975. Persistence and mobility of bromacil in orchard soils. *Weed Res.* 15:243–247.

Leonard, R.A., G.W. Langdale, and W.G. Fleming. 1979. Herbicide runoff from upland Piedmont watersheds: data and implications for modeling pesticide transport. *J. Environ. Qual.* 8:223–229.

Letey, J., and W.J. Farmer. 1974. Movement of pesticides in soil. In: W.D. Guenzi (ed.), *Pesticides in soil and water.* Soil Science Society of America. Madison. pp. 67–87.

Lindstrom, F.T., R. Haque, V.H. Freed, and L. Boersma. 1967. Theory on the movement of some herbicides in soils. *Environ. Sci. Tech.* 1:561–565.

Liu, G.C., and G.T. Felbeck, Jr. 1972. Atrazine hydrolysis as catalyzed by humic acids. *Soil Sci.* 114:201–209.

Mabey, W., and T. Mill. 1978. Critical review of hydrolysis of organic compounds in water under environmental conditions. *J. Phys. Chem. Ref. Data.* 7:383–402.

Marriage, P.B., S.U. Khan, and W.J. Saidak. 1977. Resistence and movement of terbacil in peach orchard soil after repeated annual applications. *Weed Res.* 17:219–255.

Matsunaka, S., and S. Kuwatsuka. 1975. Environmental problems related to herbicide use in Japan. *Environ. Qual. Safety* 4:149–159.

McClure, G.W. 1970. Accelerated degradation of herbicides in soil by the application of microbial nutrient broths.. *Contn. Boyce Thompson Inst. Plant Res.* 24:235–240.

McCormick, L.L., and A.E. Hiltbold. 1966. Microbiological decomposition of atrazine and diuron in soil. *Weeds* 14:77–82.

McDonald, K.B., R.B. McKercher, and J.R. Moyer. 1976. Picloram displacement in soil. *Soil Sci.* 121:94–102.

McKay, D., and P.J. Leinonen. 1975. Rate of evaporation of low-solubility containments from water bodies to atmosphere. *Environ. Sci. Tech.* 9:1178–1180.

Meikle, R.W., C.R. Youngson, R.T. Hedlung, C.A.I. Goring, J.W. Hamaker, and W.W. Addington. 1973. Measurement and prediction of picloram disappearance rates from soil. *Weed Sci.* 21:549–555.

Menkveld, B., and J.H. Dekker. 1983. Accelerated breakdown of butylate+ in soils with a history of its use. *North Cent. Weed Cont. Conf. Proc.* 38:46.

Merkle, M.G., and R.W. Bovey. 1975. *Movement of pesticides in surface water.* Texas A&M University, College Station, TX.

Miaullis, B., G.J. Nohynek, and F. Pereino. 1982. R-33868: a novel concept for extended weed control by thiolocarbamate herbicides. *Proc. 1982 Brit. Crop. Prot. Conf. Weeds.* pp. 205–210.

Mielke, L.N. and J.R.C. Leavitt. 1981. Herbicide loss in runoff water and sediment as affected by center pivot irrigation and tillage treatments. Project completion report, Project A-062-NEB, Office of Water Research and Technology, U.S. Dept. of Interior, Washington, D.C.

Mingelgrin, U., and Z. Gerstl. 1983. Reevaluation of partitioning as a mechanism of nonionic chemicals adsorption in soils. *J. Environ. Qual.* 12:1–11.

Nakamura, Y., K. Ishikawa, and S. Kuwatsuka. 1977. Degradation of benthiocarb (thiobenicarb) in soils as affected by soil conditions. *J. Pest. Sci.* 2:7–16.

Nielsen, D.R., P.J. Wierenga, and J.W. Biggar. 1983. Spatial soil variability and mass transfers from agricultural soils. In: *Chemical mobility and reactivity in soil systems.* Soil Sci. Soc. Amer. Spec. Publ. No. 11. Soil Science Society of America, Madison. pp. 65–78.

Nilles, G.P., and M.J. Zabik. 1974. Photochemistry of bioactive compounds. Multiphase decomposition of basalin. *J. Agr. Food Chem.* 27:684–688.

Nilles, G.P., and M.J. Zabnik. 1975. Photochemistry of bioactive compounds. Multiphase photodegradation and mass spectral analysis of basagran. *J. Agr. Food Chem.* 23:410–415.

Nygen, L., and A. Lange. 1978. A comparison of 6 herbicides applied through drip irrigation on indicator crops. *West. Soc. Weed Sci. Res. Prog. Rep.* pp. 62–63.

O'Conner, G.A., and J.U. Anderson. 1974. Soil factors affecting the adsorption of 2,4,5-T. *Soil Sci. Soc. Am. Proc.* 38:433–436.

Ogg, A.G., Jr. 1980. A sprinkler system for research on applying herbicides in irrigation water. *Weed Sci.*. 28:201–203.

Ogle, R.E., and F.G. Warren. 1954. Fate and activity of herbicides in soils. *Weeds* 3:257–273.

Olsen, S.R., and W.O. Kemper. 1968. Movement of nutrients to plant roots. *Adv. Agron.*. 20: 91–151.

Parker, L.W., and K.G. Doxtader. 1983. Kinetics of the microbial degradation of

2,4-D in soil: effects of temperature and moisture. *J. Environ. Qual.* 12:553–558.

Parmele, L.H., E.R. Lemon, and A.W. Taylor. 1972. Micrometerorological measurement of pesticide vapor flux from bare soil and corn under field conditions. *Water, Air, and Soil Pollut.* 1:433–451.

Peterson, M.A., and W.E. Arnold. 1983. Effect of soil type on chlorsulfuron carryover in Eastern South Dakota. *North Cent. Weed Cont. Conf.* 38:35.

Philen, O.D., Jr., S.B. Weed, and J.B. Weber. 1970. Estimation of surface charge density of mica and vermiculite by competitive adsorption of diquat vs. paraquat. *Soil Sci. Soc. Am. Proc.* 34:527–531.

Philen, O.D., Jr., S.B. Weed, and J.B. Weber. 1971. Surface charge characteristics of layer silicates by competitive adsorption of two organic divalent cations. *Clays Clay Miner.* 19:259–302.

Plimmer, J.R. 1976. Volatility. In: P.C. Kearney and D.D. Kaufmann (eds.), *Herbicides: chemistry, degradation and mode of action. Vol. II.* Marcel Dekker, New York. pp. 891–934.

Probst, G.W., and J.B. Tepe. 1969. Trifluralin and related compounds. In: P.C. Kearney and D.D. Kaufman (eds.), *The degradation of herbicides.* Marcel Dekker, New York. pp. 255–282.

Probst, G.W., T. Golab, and W.L. Wright. 1975. Degradation of herbicides. In: P.C. Kearney and D.D. Kaufman (eds.), *The degradation of herbicides.* Marcel Dekker , New York. p. 453.

Probst, G.W., T. Golab, R.J. Herberg, F.J. Holzer, S.J. Parka, C. van der Schans, and J.B. Tepe. 1967. Fate of trifluralin in soils and plants. *J. Agr. Food Chem.* 15:592–599.

Raats, P.A.C. 1984. Tracing parcels of water and solutes in unsaturated zones. In: B. Yaron, G. Dayan and J. Goldschmidt (eds.), *Pollutants in porous media.* Springer-Verlag, Heidelberg. pp. 4–16.

Raney, W.A. 1965. Physical factors of the soil as they affect microorganisms. In: K.F. Baker and W.C. Snyder (eds.), *Ecology of soil-borne plant pathogens.* University of California Press, Berkeley. pp. 115–119.

Rao, P.S.C., and J.M. Davidson. 1980. Estimation of pesticide retention and transformation parameters required in nonpoint source pollution models. In: M.R. Overcash (ed.), *Environmental impact of non-point source pollution.* Ann Arbor Science Publishers, Ann Arbor, MI. pp. 23–67.

Rao, P.S.C., and R.E. Jessup 1982. Development and verification of simulation models for describing pesticide dynamics in soils. *Ecol. Model.* 16:67–76.

Rao, P.S.C., and R.E. Jessup. 1983. Sorption and movement of pesticides and other toxic organic substances in soils. In: *Chemical mobility and reactivity in soil systems.* Soil Sci. Soc. Am. Spec. Publ. No. 11. Soil Science Society of America, Madison. pp. 183–201.

Rao, P.S.C., J.M. Davidson, R.E. Jessup, and H.M. Selim. 1979. Evaluation of conceptual models for describing nonequilibrium adsorption-desorption of pesticides during steady-flow in soils. *Soil Sci. Soc. Am. J.* 43:22–28.

Rhodes, R.C., I.J. Belasco, and H.L. Pease. 1970. Determination of mobility and adsorption of agrichemicals on soils. *J. Agr. Food Chem.* 18:524–528.

Rudyanski, W.J., and R.S. Fawcett. 1983. Effect of herbicide and insecticide use history of thiocarbamate herbicide persistence. *North Cent. Weed Cont. Conf. Proc.* 38:39.

Russell, J.D., M. Guy, J.L. White, G.W. Bailey, W.R. Payne, J.O. Pope, and J.I. Teasley. 1968. Model of chemical degradation of triazines by montmorillonite. *Science* 160:1340–1342.

Saltzman, S., Z. Gerstl, and B. Yaron. 1981. Factors to be considered in the application of pesticides via drip irrigati on. In: *Colloque Franco-Israelien: Les phenomenes de transport de l'eau et des solutes et l'irrigation.* Auignon, France. pp. 71–82.

Savage, K.E. 1977. Metribuzin persistence in soil. *Weed Sci.* 25:55–59.

Savage, K.E., and R.D. Wauchope. 1974. Fluometuron adsorption-desorption equilibria in soil. *Weed Sci.* 22:106–110.

Schiavon, M., and F. Jacquin. 1973a. Contribution a l'etude de la migration de deux triazines sans l'influence des precipitations. *Proc. Eur. Weed Res. Coun. Symp. Herb. Soil.* pp. 80–96.

Schiavon, M., and F. Jacquin. 1973b. Etude de la presence d'atrazine dans les eaux de drainage. *C.R. Journees d'Etudes Herbicides.* Columna 35–43.

Schnitzer, M., and S.U. Khan. 1972. *Humic substances in the environment.* Marcel Dekker, New York.

Scott, H.D., and R.E. Phillips. 1972. Diffusion of selected herbicides in soil. *Soil Sci. Soc. Am. Proc.* 36:714–719.

Scott, H.D., and R.E. Phillips. 1973. Movement of herbicides through soil to soybean roots. *Argon. J.* 65:386–390.

Scow, K.M. 1982. Rate of biodegradation. In: W.J. Lyman, W.F. Reehl, and D.H. Rosenblatt (eds.), *Handbook of chemical property estimation methods.* McGraw-Hill Co., New York. pp. 9-1–9-85.

Sheets, T.J., and C.I. Harris. 1965. Herbicide residues in soils and their phytotoxicity to crops grown in rotation. *Res. Rev.* 11:119–140.

Skipper, H.D., C.M. Gilmour, and W.R. Furtick. 1967. Microbial versus chemical degradation of atrazine in soils. *Soil Sci. Soc. Am. Proc.* 31:653–656.

Slade, P. 1966. The fate of paraquat applied to plants. *Weeds* 6:158–167.

Smith, D.T., R.C. Berner, and J.P. Walter. 1973. Nitralin and trifluralin incorporation by rainfall and irrigation. *Weed Res.* 13:359–366.

Soderquist, C.I., J.B. Bowers, and D.G. Crosby. 1977. Dissipation of molinate in a rice field. *J. Agr. Food Chem.* 25:940–945.

Spencer, W.F. In press. Pesticides in surface irrigation runoff water. In: Natural resources modeling symposium proceedings. *USDA Cons. Res. Rep.* No. 31.

Spencer, W.F., and M.M. Cliath. 1973. Pesticide volatilization as related to water loss from soil. *J. Environ. Qual.* 2:284–289.

Spencer, W.F., and M.M. Cliath. 1974. Factors affecting vapor loss of trifluralin from soil. *J. Agr. Food Chem.* 22: 987–991.

Spencer, W.F., M.M. Cliath, and W.J. Farmer. 1969. Vapor density of soil-applied dieldrin as related to soil-water content, temperature, and dieldrin concentration. *Soil Sci. Soc. Am. Proc.* 33:509–511.

Spencer, W.F., W.J. Farmer, and M.M. Cliath. 1973. Pesticide volatilization. *Res. Rev.* 49:1–47.

Spencer, W.F., W.J. Farmer, and W.J. Jury. 1982. Behavior of organic chemicals at soil, air, and water interfaces as related to predicting the transport and volatilization of organic pollutants. *Environ. Toxic. Chem.* 1:17–26.

Spencer, W.F., T.D. Shoup, and R.C. Spear. 1980. Conversion of parathion to paraoxon on soil dusts as related to atmospheric oxidants at three California locations. *J. Agr. Food Chem.* 28:1295–1300.

Spencer, W.F., M.M. Cliath, J. Blair, and R.A. LeMert. In press. Transport of pesticides from irrigated fields in surface runoff and tile drain water. *USDA Cons. Res. Rep.* No. 31.

Stevenson, F.J. 1972. Role and function of humus in soil with emphasis on adsorption of herbicides and chelation of micronutrients. *BioScience* 22:643–650.

Stevenson, F.J. 1982. *Humus chemistry.* John Wiley and Sons, New York. 443 p.

Taylor, A.W. 1978. Post-application volatilization of pesticides under field conditions. *J. Air Poll. Contr. Assoc.* 28:922–927.

Theng, B.K.G. 1974. *The chemistry of clay-organic reactions.* Adam Hilger, London. 343 p.

The Pesticide Manual—A World Compendium, 1983. 7th Edition. C.R. Worthing (ed.). *The British Crop Protection Council.* p. 695.

Thomas, V.M., and G.L. Holt. 1980. The degradation of [$^{14}$C] molinate in soil under flooded and nonflooded conditions. *J. Environ. Sci. Health* B15:475–484.

Turner, B.C., D.E. Glotfelty, A.W., Taylor, and D.R. Watson. 1978. Volatilization of microencapsulated and conventionally applied chlorpropham in the field. *Agron. J.* 70:933–937.

Tuxhorn, G.L., F.W. Roeth, and A.R. Martin. 1983. Butylate persistence in soils with butylate history. *North Cent. Weed Cont. Conf. Proc.* 38:38.

van Genuchten, M.T., and P.J. Wierenga. 1976. Mass transfer studies in sorbing porous media. I. Analytical solutions. *Soil Sci. Soc. Am. Proc.* 40:473–480.

van Genuchten, M.T., M. Davidson, and P.J. Wierenga. 1974. An evaluation of kinetic and equilibrium equations for the prediction of pesticide movement through porous media. *Soil Sci. Soc. Am. Proc.* 38:29–34.

van Genuchten, M.T., P.J. Wierenga, and G.A. O'Connor. 1977. Mass transfer in sorbing porous media. III. Experimental evaluation with 2,4,5-T. *Soil Sci. Soc. Am. J.* 41:278–285.

van Valkenburg, W. 1973. *Pesticide formulations.* Marcel Dekker, New York.

Vinten, A.J.A., B. Yaron, and P.H. Nye. 1983. Vertical transport of pesticides into soil when adsorbed on suspended particles. *J. Agr. Food Chem.* 31:662–664.

Walker, A. 1974. A simulation model for prediction of herbicides persistence. *J. Environ. Qual.* 3:396–401.

Walker, A., and A. Barnes. 1981. Simulation of herbicide persistence in soil; a revised computer model. *Pest. Sci.* 12:123–132.

Walker, A., and P.A. Brown. 1982. Crop responses to low doses of pendimethalin, napropamide, metazachlor, and chlorsulfuron in the soil. *Proc. 1982 Brit. Crop Prot. Conf. Weeds.* pp. 141–147.

Walker, A., and D.V. Crawford. 1968. The role of organic matter in adsorption of the triazine herbicides by soils. Isotopes and radiation in soil organic matter studies. *Proc. 2nd Symp. Inst. Atom. Energy Agency* (Vienna). pp. 91–108.

Walker, A., and D.V. Crawford. 1970. Diffusion coefficients for two triazines herbicides in six soils. *Weed Res.* 10:126–132.

Walker, A., and J.A. Thompson. 1977. The degradation of simazine, linuron and propysamide in different soils. *Weed Res.* 17:399–405.

Walker, A., and R.L. Zimdahl. 1981. Simulation of the persistence of atrazine linuron and metalachlor in soil at different sites in U.S.A. *Weed Res.* 21:255–265.

Walker, A., R.J. Hance, J.G. Allen, G.G. Briggs, Y.L. Chen, J.D. Gaylor, J. Hogue, A. Malquou, K. Moody, R. Moyer, W. Pestemer, A. Rahman, A.E. Smithe, J.C. Streibig, N.T.L. Torstensson, L.S. Widyanto, and Z. Zandvoort. 1983. EWRS herbicide-soil working group: collaborative experiment on simazine persistence in soil. *Weed Res.* 23:373–383.

Watson, J.R., A.M. Posner, and J.P. Quirk. 1973. Adsorption of the herbicide 2,4-D on a soil mineral goethite. *J. Soil Sci.* 24:503–511.

Wauchope, R.D. 1975. Fixation of arsenical herbicides, phosphate, and arsenate in alluvial soils. *J. Environ. Qual.* 4:355–358.

Wauchope, R.D. 1978. The pesticide content of surface water draining from agricultural fields—a review. *J. Environ. Qual.* 7:459–472.

Weber, J.B. 1967. Spectrophotometrically determined ionization constants of 13 alkylamino *s*-triazines and the relationships of molecular structure and basicity. *Spectrochimica Acta* 23:458–461.

Weber, J.B. 1970. Adsorption of *s*-triazine by montmorillonite as a function of pH and molecular structure. *Soil Sci. Soc. Am. Proc.* 34:401–404.

Weber, J.B., and S.B. Weed. 1968. Adsorption and desorption of diquat, paraquat, and prometone by montmorillonitic and kaolinitic clay minerals. *Soil Sci. Soc. Am. Proc.* 32:485–487.

Weber, J.B., and D.M. Whitacre. 1982. Mobility of herbicides in soil columns under saturated- and unsaturated-flow conditions. *Weed Sci.* 30:579–584.

Wheatley, G.A. 1973. Pesticides in the atmosphere. In: C.A. Edwards (ed.), *Environmental pollution by pesticides*. Plenum Press, London. pp. 365–408.

White, A.W., Jr., L.A. Harper, R.A. Leonard, and J.W. Turnbull. 1977. Trifluralin volatilization losses from a soybean field. *J. Environ. Qual.* 6:105–110.

Wiese, A.F., and W.E. Turner. 1978. Herbicide application with sprinkler irrigation. *Proc. South. Weed Sci. Soc.* 31:100.

Williams, P.P. 1977. Metabolism of synthetic organic pesticides by anaerobic microorganisms. *Res. Rev.* 66:63–135.

Wolf, D.E., and K.P. Martin. 1974. Microbial degradation of 2-carbon$^{-14}$ bromacil and terbacil. *Soil Sci. Soc. Am. Proc.* 38:921–925.

Wu, C.H., N. Buehring, J.M. Davidson, and P.W. Santelman. 1975. Napropamide adsorption, desorption, and movement in soils. *Weed Sci.* 23:454–457.

Young, J.C., and S.U. Khan. 1978. Kinetics of nitrosation of the herbicide glyphosate. *J. Environ. Sci. Health* B13:59–72.

Zepp, R.G., N.L. Wolfe, G.L. Baughman, and R.C. Hollis. 1977. Singlet oxygen in natural waters. *Nature* 267:421–423.

Zimdahl, R.L., and W.A. Fithian. 1984. Chlorsulfuron soil persistence. *West. Soc. Weed Sci. Res. Prog. Rep.* pp. 278–281.

Zurawski, H., and J.S. Piss. 1971. Laboratory investigations on the rate of inactivation of simazine. *Reczniki Glebozn.* pp. 127–136. (*Weed Abstracts* No. 245, 1973.)

# Index

## R



## S



## T

Made in the USA
Monee, IL
18 October 2026

66036448R10129